ESG Factors and Financial Outcomes in Banks

Nicola Del Sarto

ESG Factors and Financial Outcomes in Banks

Quantitative Insights into the ESG-Cost of Debt
Relationship

Nicola Del Sarto
University of Florence
Florence, Italy

ISBN 978-3-031-87747-6   ISBN 978-3-031-87748-3   (eBook)
https://doi.org/10.1007/978-3-031-87748-3

This Palgrave Macmillan imprint is published by the registered company Springer Nature Switzerland AG
The registered company address is: Gewerbestrasse 11, 6330 Cham, Switzerland

If disposing of this product, please recycle the paper.

To my family…

Introduction

Sustainability has become a crucial issue for today's economy, with the financial sector at the forefront of this shift. The rising emphasis on environmental, social, and economic sustainability stems from a recognition that addressing these issues is not just a moral or environmental concern but essential for building long-term economic resilience and societal well-being (Clark et al., 2015). Recent years have seen intensified attention on phenomena like climate change, pollution, resource depletion, and social inequality, each posing direct threats to economies and communities. What were once considered abstract concepts are now material realities that demand immediate attention from governments, financial institutions, and market actors worldwide. This growing awareness has catalyzed a paradigm shift, where sustainability has become central to economic planning and corporate strategy, redefining the goals and scope of financial markets (Schoenmaker & Schramade, 2019).

The shift toward sustainability also requires a reassessment of traditional economic models, particularly with the integration of frameworks like ESG (Environmental, Social, and Governance) criteria. Since their introduction in the early 2000s, ESG factors have redefined what constitutes corporate success, shifting the focus from short-term financial gains to long-term stability and resilience (Eccles et al., 2014). ESG criteria have become essential in evaluating companies, moving beyond conventional financial metrics to include non-financial risks and opportunities that impact a company's performance, reputation, and viability. Companies embracing ESG are increasingly seen as forward-thinking, capable of mitigating risks associated with environmental degradation, social unrest, or governance failures. Research shows that ESG-oriented companies tend to exhibit stronger financial performance, resilience, and competitiveness (Friede et al., 2015), highlighting that sustainable practices are not just ethical but economically beneficial.

However, despite the growing prominence of ESG criteria, significant challenges remain, particularly regarding the standardization of ESG metrics. The lack of universally accepted definitions and measurement frameworks creates inconsistencies in how ESG performance is assessed, reported, and interpreted. These disparities complicate efforts to compare companies across industries or regions and may lead to issues like greenwashing, where firms exaggerate or misrepresent their sustainability efforts. For financial institutions and researchers, this lack of standardization poses a challenge in accurately integrating ESG considerations into risk assessments, investment decisions, and lending practices. Addressing these issues is critical to ensuring the credibility and effectiveness of ESG as a tool for driving sustainable transformation.

This book seeks to explore this transformative shift in both economic and cultural contexts, focusing on ESG's growing role in corporate finance and bank lending. The empirical analysis in this book investigates whether companies that actively engage in ESG practices experience reduced costs of debt financing, a question that has received limited attention in the literature, as most research has focused on the cost of equity or CSR impacts rather than debt financing (Goss & Roberts, 2011). This study fills this gap by examining whether banks perceive ESG-aligned companies as lower-risk borrowers, potentially rewarding them with more favorable loan terms. Such insights are crucial, as banks play a key role in capital allocation, influencing companies' access to financing and driving sustainable growth.

The central hypothesis of this research posits a negative relationship between ESG performance and the cost of debt, suggesting that companies committed to ESG may be perceived as less risky and receive more favorable borrowing terms. This aligns with a broader trend where financial markets increasingly recognize that sustainability is not only a corporate responsibility but also a strategic advantage that can mitigate risk and enhance long-term returns.

The book is organized into the following sections:

Chapter 1 explores the historical evolution of sustainability concepts and sustainable finance, from early ideas on corporate social responsibility (CSR) to the rise of ESG criteria. This chapter reviews key management theories, such as stakeholder theory and shared value creation, which have underpinned the shift towards sustainable practices. It also examines the significant role of banks in promoting ESG initiatives by embedding sustainability into their lending strategies.

Chapter 2 delves into the regulatory landscape shaping sustainable finance, particularly in the European Union. It highlights major initiatives like the European Green Deal, the Sustainable Finance Disclosure Regulation (SFDR), and the EU Taxonomy for sustainable finance. This chapter emphasizes the role of regulation in guiding financial markets toward sustainable practices, setting the foundation for future policy shifts aligned with environmental, social, and governance goals. By examining these frameworks, the chapter illustrates how regulatory advancements aim to create a cohesive approach to sustainability across financial markets.

Chapter 3 presents the book's empirical analysis, focusing on data from 107 European companies between 2017 and 2021. The analysis investigates the potential link between a company's ESG performance and its cost of debt, testing the hypothesis that banks may offer more favorable lending rates to ESG-aligned companies. The chapter also contextualizes these findings within the EU's broader commitment to sustainability, providing insights into how banks might leverage ESG considerations to foster responsible corporate behavior and create financial incentives for sustainability.

Chapter 4 Future Directions and Innovations in Sustainable Finance: Challenges and Opportunities looks ahead to the future of sustainable finance, discussing emerging trends, technological innovations, and unresolved challenges in ESG integration. This chapter examines the role of artificial intelligence, blockchain, and other digital tools in improving ESG transparency and standardization, as well as the potential for these technologies to address greenwashing concerns. It also highlights the need for global harmonization of ESG metrics and explores innovative financing mechanisms, such as green bonds and sustainability-linked loans, that could drive further progress in aligning financial markets with sustainability goals. In summary, this book contributes to the growing body of literature on sustainable finance by exploring how ESG factors influence corporate financing,

particularly regarding the cost of debt. By examining the interplay between ESG and the financial sector, the book highlights
the evolving role of banks in driving the transition to a sustainable economy. The findings underscore the significance of regulatory
progress, ESG integration in investment and lending practices, and
the financial sector's capacity to support sustainable development
through responsible capital allocation. This work ultimately reflects
the broader economic transformation underway as finance moves to
support a resilient, inclusive, and sustainable future.

CONTENTS

LIST OF FIGURES

List of Tables

CHAPTER 1

The Evolution of Sustainability: From CSR to Sustainable Finance

Abstract This chapter traces the evolution of sustainable finance, from early theories like the Malthusian trap to modern frameworks such as the Brundtland Report and the UN's 2030 Agenda. Key milestones, including the Stockholm Conference and The Limits to Growth report, highlight the tension between economic growth and resource limitations. The chapter examines global progress using tools like the Environmental Performance Index (EPI) and initiatives such as the European Green Deal. It emphasizes the critical need for coordinated actions to balance environmental, social, and economic dimensions for a sustainable future.

Keywords Sustainable finance · Environmental sustainability · Malthusian trap · Brundtland Report · Environmental Performance Index (EPI)

THE EVOLUTION OF SUSTAINABLE FINANCE

The growing focus on sustainability-related issues has become increasingly pressing, as challenges like climate change, pollution, ozone depletion, deforestation, and resource degradation pose urgent threats to the planet's future (Rockström et al., 2009). The increasing awareness of these issues and the need for sustainable action reflect a journey

© The Author(s), under exclusive license to Springer Nature Switzerland AG 2025
N. Del Sarto, *ESG Factors and Financial Outcomes in Banks*, https://doi.org/10.1007/978-3-031-87748-3_1

1

rooted in historical thought and scientific inquiry. As early as 1798, economist Thomas Malthus proposed the concept of the "Malthusian trap," warning of an impending imbalance between population growth and food resources. While his predictions were later challenged, Malthus's work introduced a foundational tension between economic growth and resource limitations that remains pertinent in today's sustainability debates (Jackson & Webster, 2016; Meadows et al., 1972).

This conversation took a data-driven turn in the 1970s, notably with the Club of Rome's report The Limits to Growth, which was commissioned by MIT's System Dynamics Group to create a global model of the human impact on five key variables: population, food production, industrialization, pollution, and resource depletion. Using systems modeling, this report projected that exponential growth in human activity could exceed the Earth's finite resources, potentially leading to societal collapse unless a balance was achieved between human demands and environmental limits. The report was groundbreaking, sparking a wave of awareness among scientists, policymakers, and the public regarding the long-term environmental consequences of unchecked development (Meadows et al., 1972; Turner, 2008). It served as a foundational text in the field of environmental science and encouraged further exploration of sustainable practices (Bardi, 2011).

The oil shocks of the 1970s highlighted the fragility of reliance on finite resources, particularly fossil fuels, and brought energy security to the forefront of global concerns (Stern, 2012). The surge in oil prices and ensuing economic instability catalyzed an interest in renewable energy sources and prompted discussions on the limitations of the Western industrial model, which was built on resource-intensive production. The global recognition of these vulnerabilities led the United Nations to convene the Stockholm Conference in 1972, marking the first international gathering to address the interconnected issues of environment and development. The Stockholm Declaration that emerged from this conference articulated the essential need for international cooperation to protect the environment and introduced the concept of environmental rights and responsibilities, highlighting the duty of present generations to preserve resources for the future (UNEP, 2019). This conference also established the United Nations Environment Programme (UNEP), which continues to play a pivotal role in promoting environmental sustainability worldwide.

The events of this period underscored the unsustainability of unchecked economic growth and placed human behavior at the center of environmental challenges, leading to a paradigm shift in how governments and societies viewed development. In 1987, sustainable development was officially defined by the Brundtland Commission as "development that meets the needs of the present without compromising the ability of future generations to meet their own needs," setting a new standard for global development goals (Brundtland, 1987). This definition broadened the scope of sustainability beyond environmental concerns, integrating social and economic considerations and establishing a framework for future global policy (WCED, 1987).

Today, the importance of sustainability has only grown, and countries vary widely in their approaches and successes in addressing these issues. The Environmental Performance Index (EPI), developed by Yale and Columbia universities, ranks countries based on metrics such as climate health, pollution control, and ecosystem vitality. In 2022, Denmark, the United Kingdom, and Finland topped the EPI rankings, highlighting their commitment to climate action and ecosystem preservation. Meanwhile, larger economies like the U.S. and China, which face challenges in emissions reductions and ecosystem protection, lagged in comparison (EPI, 2022). The disparities in these rankings illustrate the varied approaches taken by global powers and emphasize the urgency for cohesive, scalable strategies to address environmental, social, and economic sustainability (Hickel & Kallis, 2020).

This ongoing global transition toward sustainability continues to be driven by advancements in policy, science, and social awareness. Recent initiatives, such as the European Green Deal and the UN's 2030 Agenda, demonstrate a renewed commitment to addressing sustainability in a coordinated and structured manner (European Commission, 2019). While significant progress has been made, these efforts also reveal the challenges of implementing sustainable practices in diverse political and economic contexts. Together, these historical milestones and current initiatives underscore the pressing need for transformative actions that integrate environmental stewardship, social equity, and economic resilience into the core of global development (Fig. 1.1).

The Environmental Performance Index (EPI) report highlights that the wealthiest democracies consistently rank at the top, as is often the case in other environmental assessments. The countries that have achieved strong results demonstrate a deep commitment to all areas of

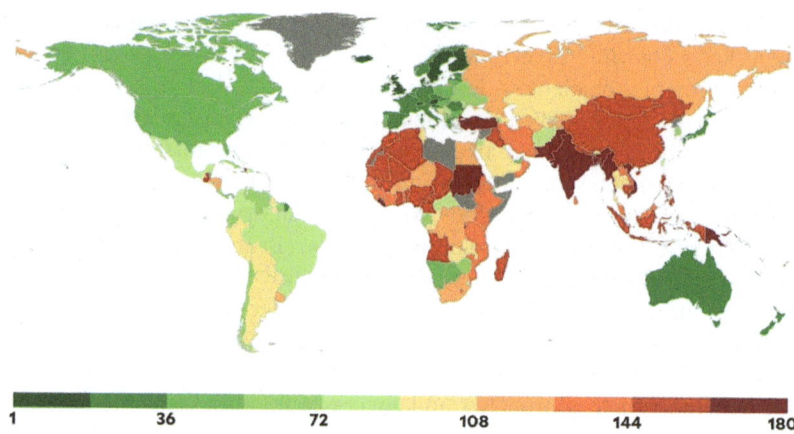

Fig. 1.1 Ranking of the 2022 Environmental Performance Index (EPI) for 180 countries (*Source* EPI 2022. Ranking country performance on sustainability issues, p. 16)

sustainability, supporting their policy goals with stringent regulations and financial investments. However, even the "greenest" countries still have room for improvement. These high-performing nations can also serve as role models, sharing their policy practices with those countries that have fallen behind in the pursuit of a sustainable future. Broadly speaking, many leading countries in environmental health show weaker performance in addressing climate change. Ecosystem vitality results are also inconsistent, reflecting the need for increased investments in decarbonization, biodiversity protection, and habitat preservation worldwide.

As in 2020, Denmark ranks first (highlighted in dark green) due to its global leadership in climate and sustainable agriculture. This European nation has recently set a national target to reduce emissions by 70% by 2030, compared to 1990 levels, and has implemented a comprehensive political agenda to meet this commitment, including the recent extension of greenhouse gas taxes. The United Kingdom and Finland, in second and third positions respectively, stand out for their climate change performance, driven by policies in recent years that have significantly reduced greenhouse gas emissions.

The relatively low ranking of the United States is primarily due to its poor performance in combating climate change. Although greenhouse gas emissions are decreasing in the U.S., the current pace is not enough to mitigate its historically high levels. It is estimated that the U.S., along with China, India, and Russia, will be responsible for over 50% of global greenhouse gas emissions by 2050 unless climate policies and decarbonization efforts are significantly strengthened. Despite this, the U.S. has made progress in other areas of sustainability, such as air quality and marine protected areas, but the overall EPI ranking still places it behind most Western democracies.

China ranks 160th out of 180 countries, with a score of 28.4, mainly due to its limited commitment to biodiversity and ecosystem protection. However, China's ranking has improved over the past decade (+11.4 points in its overall EPI score).

At the bottom of the ranking, the countries making the least effort toward environmental sustainability are in South Asia, with the bottom five being: Pakistan (24.6 out of 100), Bangladesh (23.1), Vietnam (20.1), Myanmar (19.4), and India in last place (18.9). Broadly speaking, the Asian countries with low scores are those that have prioritized economic growth over sustainability or are grappling with severe civil unrest and crises. India, in particular, faces increasingly worrying air quality issues and a sharp rise in greenhouse gas emissions, dropping to the bottom of the ranking for the first time (it was 168th in 2020). Thus, despite recent promises from countries like India and China to curb emission growth rates, the reality painted by the data is dire. A low or very low EPI score implies that these countries must urgently review and significantly reform their national sustainability strategies, with a strong focus on decarbonization, air quality improvement, waste management, and biodiversity protection.

The EPI ranking, published on June 1st of this year, underscores that the world is far from being on track to meet its environmental sustainability commitments. The report's authors warn that the vast majority of countries will fail to achieve the goal of net-zero greenhouse gas emissions by mid-century. The analysis clearly shows that global progress in reducing greenhouse gas emissions (GHG) is still insufficient, and, at the current pace, the net-zero target by 2050—established in the 2021 Glasgow Climate Pact—will not be met. Consequently, if the world truly wants to avoid collapse and the devastating impacts of climate change, which were highlighted years ago, much more work needs to be done than has been achieved so far. Particularly alarming is the fact that in countries like China, India, Russia, and many developing nations, greenhouse gas emissions continue to rise steadily. As of now, according to EPI analysis, only a small number of countries (all in Europe) are on track to achieve greenhouse gas neutrality by meeting their commitments, with Denmark and the United Kingdom standing out in particular.

It is important to note that the EPI is regarded as a leading analysis of global sustainability trends, as it provides a powerful policy tool for evaluating a wide range of critical issues for all countries: air and water pollution, waste management, biodiversity, habitat conservation, and the transition to a clean energy future. Moreover, if we look at the factors that, according to the EPI rankings, lead to environmental success in a state, we find that they include good governance, wealth, quality of life, and well-structured regulations.

In fact, political decisions also play a crucial role in effectively managing environmental threats and guiding countries toward a more sustainable future. For these reasons, the findings of the EPI projections should be taken seriously by all political leaders and should serve as a strict reminder: "There is no more time to delay; action must be taken swiftly and concretely." Indeed, the transition toward stricter environmental policies has gained significant momentum in recent years. Consider, for instance, the United Nations' adoption of the Sustainable Development Goals (SDGs) or the signing of the Paris Agreement, both in 2015. Nevertheless, the EPI team notes that the future possibility of living in a more sustainable world is often hindered by persistent information gaps. Therefore, political efforts should also focus on better and more extensive data collection, reporting, and verification on a wide range of environmental issues (agriculture, freshwater quality, chemical exposure, ecosystem protection). Of course, this will require greater investments in environmental information systems.

As will be discussed in the following section, if the path toward sustainable development is to be pursued, there must be sufficient financial resources to invest in environmental protection.

SUSTAINABLE DEVELOPMENT

Awareness of the need for a crucial transformation in the relationship between economic activity and the natural world has led to the development of a key concept: sustainable development.

In 1983, within the framework of the United Nations, the World Commission on Environment and Development (WCED) was established. That same year, the UN Secretary-General appointed Norwegian politician Gro Harlem Brundtland as chair of the commission. Shortly thereafter, in 1987, during a WCED meeting in Stockholm, the commission—also known as the Brundtland Commission—presented its final report, Our Common Future, which introduced the now widely known concept of sustainable development:

> Sustainable development is development that meets the needs of the present without compromising the ability of future generations to meet their own needs. (WCED, Our Common Future, 1987, p. 37)

This concept ushered in a new economic model focused on qualitative improvement rather than growth as a mere quantitative increase. The Brundtland Report opened the door to a development model grounded in respect for future progress, encapsulated by the notion of "sustainability." This model was not seen as a fixed condition to aspire to, but rather as a process of change in which the use of resources, management of investments, direction of technological development, and institutional changes align with both present and future needs. The report emphasized that global environmental challenges were largely rooted in socioeconomic imbalances between wealthier countries and developing nations, stressing the importance of meeting essential needs for all individuals and advocating for a more respectful growth model. It argued that the possibility of ushering in a new era of economic growth that is environmentally sustainable must be paired with a universal legitimacy to aspire to better living conditions. Furthermore, the report reiterated the importance of international cooperation in achieving these common goals.

By highlighting the principles of intergenerational and intragenerational equity and adopting a long-term perspective, the Brundtland Report offered a serious warning:

> It is time to take the necessary decisions to ensure the resources to support this and future generations. (WCED, Our Common Future, 1987, p. 11)

This document helped introduce the concept of sustainability into global legislative frameworks. In fact, since the Brundtland Report, the concept of sustainable development has been continuously embraced and adopted as a fundamental model for new development visions.

Following the discussions surrounding this report, the United Nations General Assembly decided to organize a Conference on Environment and Development (UNCED). Held in Rio de Janeiro in 1992, this conference—also known as the Earth Summit—revisited and further elaborated on the concept of sustainable development, focusing on the impact of human socioeconomic activities on the environment. It was at this summit that the interconnectedness of social, economic, and environmental factors was highlighted, emphasizing that progress in one area requires action in all others to be sustained over time. This led to the recognition that sustainable development requires an integrated approach, balancing environmental, economic, and social dimensions.

The Rio Conference sparked a lively debate between governments and their citizens on how to ensure the sustainability of development, giving equal importance to environmental protection, economic growth, and social development. Additionally, the principle of sustainable development was formally codified in the documents adopted at the Summit.

In essence, the goal of sustainable development rests on three interrelated pillars. The first is social equity, promoting fairness between present and future generations and between the rich and poor populations of the world. The second is environmental protection, focusing on safeguarding and restoring the quality of natural resources to halt the environmental degradation caused by traditional economic development. Finally, the third pillar is economic competitiveness, defined as the ability of economic systems to generate income and employment for all, while respecting the surrounding environment.

The path toward a definitive formulation of the principles of sustainable development culminated in the creation of an important action plan for people, the planet, prosperity, peace, and partnership. This is the 2030 Agenda for Sustainable Development, a document signed in September 2015 by the governments of the 193 UN member states after years of consultations. The significance of this comprehensive action plan lies in its inclusion of 17 Sustainable Development Goals (SDGs), articulated into 169 targets to be achieved by 2030. It is important to note that the SDGs build on the achievements of the eight Millennium Development Goals (MDGs), adopted following the United Nations Millennium Declaration in 2000, with a target date of 2015. The ambitious and transformative project outlined in the 2030 Agenda seeks to create a better and more sustainable future for all by enhancing the Millennium Goals and addressing those that were not fully achieved. Specifically, the 17 Sustainable Development Goals represent common objectives for all countries and individuals in addressing global challenges, including poverty, inequality, climate change, environmental degradation, justice, and peace.

Furthermore, the 17 SDGs are all interconnected, balancing the three aspects of sustainable development: economic, social, and environmental. As stated on page 6 of the resolution adopted by the General Assembly on September 25, 2015—the 2030 Agenda for Sustainable Development, titled Transforming Our World: The 2030 Agenda for Sustainable Development:

> Today, we are announcing 17 new Sustainable Development Goals with 169 associated targets, which are integrated and indivisible. This is the first time that world leaders have committed to a shared effort and action through such a broad and universal political agenda. We are embarking on a path toward sustainable development, dedicating ourselves to pursuing global growth and cooperation that will result in greater benefits for all countries and the entire world. We reaffirm that all states can and must freely exercise their full and permanent sovereignty over their wealth, natural resources, and economic activities. We will implement the Agenda so that all can benefit, for the present and future generations.

Thus, every country in the world must collaborate to achieve the Sustainable Development Goals, committing to implementing this ambitious program and monitoring its progress. The breakdown of the 17 Sustainable Development Goals outlined in the 2030 Agenda is as follows and is summarized in Fig. 1.2.

The Sustainable Development Goals (SDGs) and their associated targets came into effect on January 1, 2016, guiding the decisions of all countries at regional and global levels for the next 15 years. Another significant event of 2015 was the Paris Agreement on Climate Change, the first universal accord—signed by the member states of the UN Framework Convention on Climate Change (UNFCCC)—to combat global warming by reducing carbon dioxide emissions. From the discussion in this section, it is clear that the ideas of environmental sustainability and sustainable development have spread worldwide, thanks to international resolutions and the resulting documents. Examples include the 2030 Agenda, the SDGs, and the Paris Climate Agreement (COP21).

Fig. 1.2 The 17 sustainable development goals (*Source* United Nations)

However, when considering the environmental dimension of sustainability, it is the European Union that first implemented legislation to reduce greenhouse gas emissions and align with the 2015 Paris Agreement. The EU and its member states have committed to becoming the world's first climate-neutral economy by 2050. The EU is renowned for being a global leader in environmental policy, especially compared to other major countries like the United States, China, Russia, and India. Through its leadership and ambitious sustainability policies, the EU has become a model that can potentially inspire and encourage other world leaders to promote sustainable development for the planet.

One example of this is the European Green Deal, launched in December 2019. It is a comprehensive package of strategic initiatives implemented by the European Commission to lead the EU toward a green transition and achieve climate neutrality by 2050. The Green Deal represents the EU's strategy to meet the goals set out in the 2015 Paris Agreement. It encompasses concrete actions and policies across all sectors of the economy, including climate, environment, energy, transport, industry, agriculture, research, innovation, and sustainable finance. Essentially, the Green Deal envisions a drastic transformation of Europe's society and economy. By combining the pressure to pursue environmental sustainability with the necessary financial resources, the European Green Deal offers a blueprint for how Europe envisions the future development of its economy and society. The hope is that through the Green Deal, the EU will accelerate the process of achieving the Sustainable Development Goals.

More recently, in June 2021, the EU adopted its first-ever climate law, which legally binds the goal of achieving climate neutrality by 2050. The same legislation mandates that the EU and its member states reduce net greenhouse gas emissions (emissions minus absorption) by at least 55% by 2030, compared to 1990 levels. The EU Climate Law (Regulation (EU) 2021/1119), also known as the "European Climate Law," came into force on July 29, 2021. It is part of the legislative reforms to implement the European Green Deal and the Union's decarbonization objectives defined by the Green Deal. This regulation integrates climate policies with sustainability policies by combining the goals of reducing greenhouse gas emissions and adapting to the impacts of climate change with sustainable development.

From the above, it is evident that the EU and its member states place climate action, sustainability, and sustainable development as top priorities. Through the European Green Deal, the EU positions itself as a global leader in climate policy and action, serving as a model for its global partners. The 27 EU countries, represented by the European Commission and the rotating presidency of the Council, play a critical role in enhancing international efforts on climate and sustainability while also providing financial support to developing countries to promote a global green transition. In this leadership role, during the most recent COP26 meeting in 2021, the EU helped urge other nations to intensify their work on climate issues and the environmental challenges the world faces today. Thus, the EU has also distinguished itself within the UN as being at the forefront of addressing environmental concerns.

Despite the significant initiatives and progress made at the close of COP26, the world still seems unprepared to fully embrace the green transition. The coming years will be critical for achieving the targets set by the Paris Agreement, but further substantial global efforts are necessary. Indeed, the EU cannot fight a global threat alone.

Simultaneously, the United States also appears determined to restore its leadership on environmental issues. Since January 20, 2021, the new U.S. President Joe Biden has taken office, immediately distancing himself from the climate disengagement of his predecessor, Donald Trump.

Under Biden's leadership, climate change has been elevated to a top priority for U.S. national security, the economy, and the country's future. One of Biden's first political acts was to rejoin the Paris Agreement on climate change, aimed at curbing global warming. His decision to re-enter the agreement—in contrast to Trump's withdrawal—offers hope for a tangible shift toward global sustainable development. Additionally, Biden has signed several executive orders banning new gas and oil extraction contracts on federal lands. The new U.S. commitment, which relies on multilateralism in international relations, is seen positively by the EU, allowing for cooperation with the American superpower on environmental issues. This U.S.-EU alliance could have greater efficacy in tackling the challenge of achieving climate neutrality by 2050 and, more broadly, in reaching the Sustainable Development Goals outlined in the 2030 Agenda. U.S. involvement in global agreements is crucial to addressing the most pressing challenge of this generation and future ones, as the country is the second-largest emitter of greenhouse gases globally, after China.

To combat the climate crisis and its impact on the environment, economy, and society, President Biden signed the Inflation Reduction Act (IRA) in August 2022. This act represents the largest climate and sustainable development spending package in U.S. history, allocating around $370 billion for climate, energy, and greenhouse gas emission reduction policies. Without delving into the details of the IRA's various measures (which also address inflation reduction and healthcare costs), studies suggest that the Inflation Reduction Act could be a decisive turning point for U.S. decarbonization, allowing the country to meet its Paris Agreement targets.

Even more recently, near the conclusion of COP26, the U.S. and China surprised the world by announcing a bilateral climate agreement. This pact between the two largest global economies—and the biggest polluters—outlines a path of cooperation to achieve the goal of limiting the global average temperature rise to no more than 1.5 °C above pre-industrial levels (as agreed upon at COP21, Paris 2015). While it is clear that environmental emergencies require global engagement and coordination, the agreement between Joe Biden (USA) and Xi Jinping (China) is a step forward from past relations. Although it lacks specific technical details, the accord signals China's growing awareness that environmental issues must be urgently addressed. While the agreement is non-binding, it offers clues about the level of commitment from the world's two largest economies toward the green transition.

For developing countries, the environmental situation remains critical. These nations need the most support and funding to embark on the path toward environmental sustainability, zero emissions, and carbon neutrality. According to the UN, developed regions must honor their collective commitment to allocate at least $100 billion annually through 2025 to help emerging regions address the climate and environmental crises. According to OECD estimates, $78.9 billion was spent on this cause in 2018, and $80.4 billion in 2019. Post-pandemic estimates seem to confirm the upward trend in funding, although the main goal has not yet been reached. This financial commitment to support developing countries in mitigating and adapting to climate change was reaffirmed at the most recent G20 meeting in Rome in 2021, where some members pledged to increase their contributions further. Nevertheless, the sustainability and environmental impact situation in emerging markets remains worrying. The latest edition of the Morningstar Sustainability Atlas reports that many emerging markets received low sustainability scores, ranking in the lowest quantiles.

This brief analysis of how the world's major economies are addressing the Sustainable Development Goals highlights the need for more substantial, concrete efforts, as the path to achieving the declared objectives remains long and steep. In conclusion, as the international community's debate on environmental issues and the various dimensions of sustainable development intensifies, the concept of sustainability itself has evolved. It now encompasses not only the environmental aspect but also the social and economic dimensions. The most comprehensive definition of sustainability considers these three dimensions simultaneously. Specifically, it refers to a state of well-being (environmental, social, and economic) and the expectation that future generations' quality of life will not be compromised. Since sustainability involves the interaction between society and the environment, it is not a static concept, as both elements are constantly co-evolving. Consequently, the concept of sustainability shifts according to the historical, territorial, socioeconomic, scientific, and technological context. What is considered sustainable today may no longer be tomorrow, and vice versa.

It is important to emphasize that, in practice, sustainability is a complex concept due to the many actors involved: every individual, as an actor in many social spheres—as a citizen, consumer, entrepreneur, or institutional representative—is impacted by sustainability issues.

The next section will focus on how these developments related to sustainability and sustainable development have been integrated into the field of economic literature.

THE SHIFT IN THE FINANCIAL WORLD: FROM TRADITIONAL TO SUSTAINABLE

Since the signing of the Paris Agreement and the establishment of the Sustainable Development Goals (SDGs), finance has taken on a central role not only in fostering economic growth but also in addressing environmental challenges and integrating social and governance factors into decision-making (Eccles & Klimenko, 2019; Schoenmaker & Schramade, 2019). This shift signals a profound transformation in economic and financial paradigms, where sustainability has become a critical aspect across all fields of knowledge. Today, corporate and financial actors increasingly recognize that sustainability requires a holistic approach—balancing economic outcomes with environmental protection and social responsibility (Clark et al., 2015).

The urgency of this shift has intensified, with finance now at the forefront of global efforts to mitigate climate change, promote social equity, and ensure robust governance structures. With climate risks posing unprecedented challenges to economies, financial institutions and investors are beginning to acknowledge the need for "transition finance," or financing aimed at helping firms shift towards lower carbon footprints. A comprehensive approach to sustainable finance goes beyond green projects, addressing sectors with traditionally high carbon emissions, like energy, construction, and heavy manufacturing, in ways that integrate new technologies and innovative practices (Caldecott, 2017).

Finance, as a discipline, is constantly evolving, adapting to and often anticipating societal shifts. Over the past decade, financial markets have increasingly placed sustainability at their core, with companies and investors embracing ESG considerations as essential to long-term value creation. This growing awareness has catalyzed a rethinking of the traditional finance paradigm, which has historically focused on short-term profit maximization and shareholder returns, often neglecting broader societal and environmental impacts (Freeman et al., 2020). Sustainable finance, grounded in ethical considerations, is becoming a strategic advantage for firms, aligning economic pursuits with a commitment to preserving natural and social resources for future generations (Elkington, 1997; Soppe, 2009).

The financial crisis of 2007–2009, triggered by the collapse of Lehman Brothers and the subprime mortgage crisis, was a wake-up call. It exposed deep structural flaws within the financial system, including a lack of transparency, ethics, and long-term accountability, particularly within the banking sector (Mauck & Salzsieder, 2018). As global economies grappled with the fallout, it became evident that a stable and resilient financial system must also be sustainable. The crisis underscored the dangers of a financial paradigm solely driven by short-term profits and shareholder primacy, which had left banks and investors vulnerable to unchecked risks and unsustainable practices (Crotty, 2009). This realization prompted a profound shift toward a finance model that incorporates sustainability principles, focusing on long-term value creation and risk management.

The demand for sustainability in finance has spurred the development of sustainable finance as a distinct paradigm, characterized by the integration of ESG factors into investment and lending decisions. This shift is supported by evidence that companies with strong ESG performance tend to achieve better long-term results, both financially and in terms of stakeholder engagement (Friede et al., 2015; Khan et al., 2016). Sustainable finance thus represents a broader, three-dimensional approach, where decisions consider economic, social, and environmental impacts. In the wake of the 2007–2009 crisis, sustainable finance has focused on addressing environmental risks, promoting social equity, and improving governance, emphasizing ESG elements as drivers of risk mitigation and financial performance. Importantly, this new model sees the financial sector as a key player in the sustainability transformation, capable of using capital allocation to influence corporate behavior, reward responsible practices, and drive systemic change (Heinkel et al., 2001).

Sustainable finance has been defined in various ways, with a widely accepted definition as the management of ESG impacts within financial services, balancing profitability with broader ethical and environmental objectives (Schoenmaker & Schramade, 2019). Soppe (2009) encapsulates this perspective by describing sustainable finance as an integrated, long-term approach where financial decisions aim to optimize not only economic outcomes but also social and environmental missions: "Sustainable finance involves institutional policies, or analytical systems, where all financial decisions aim for an integrated long-term approach to optimize a company's social, environmental, and financial mission." This reflects a commitment to balancing traditional financial goals with sustainable value creation, thus redefining finance as a service to society rather than an end in itself.

In its broadest sense, sustainable finance applies the principles of sustainable development to financial activities, integrating environmental and social considerations into long-term investment strategies. This approach necessitates systemic changes across the financial landscape, affecting households, businesses, governments, investors, regulators, and supervisory bodies, and calls for the incorporation of non-financial factors into decision-making processes (Clark et al., 2018). Sustainable finance shifts the objective of finance from pure profit maximization to serving societal goals, supporting the urgent transition to a low-carbon, inclusive economy (Schoenmaker, 2017). By aligning financial activities with sustainable development, finance can be a powerful tool to drive the necessary transformation towards a more resilient and equitable global economy.

The growing adoption of sustainable finance principles has prompted significant regulatory and policy actions, such as the EU's Action Plan on Sustainable Finance and the UN's Sustainable Development Goals, which seek to build robust frameworks to ensure transparency and accountability in sustainable investment practices. Such initiatives provide guidance for integrating ESG considerations into the financial system, further embedding the principle that economic growth should harmonize with ecological stewardship and social well-being (European Commission, 2018; United Nations, 2015). This underscores that sustainable finance is not just a response to environmental crises but a transformative movement reshaping finance's role in addressing the complex and interrelated issues facing society today.

In summary, the shift from traditional finance to sustainable finance, as illustrated in Fig. 1.3, highlights a growing preference for models that integrate environmental, social, and governance (ESG) concerns into the financial decision-making process. Unlike traditional finance, which relies on a linear production and consumption model focused on short-term financial gain, sustainable finance adopts a circular approach that emphasizes preserving financial, natural, and social capital over the long term. This shift marks a critical transformation in value creation, underscoring an integrated, three-dimensional approach that balances economic returns with social and environmental outcomes (Clark et al., 2015; Schoenmaker & Schramade, 2019).

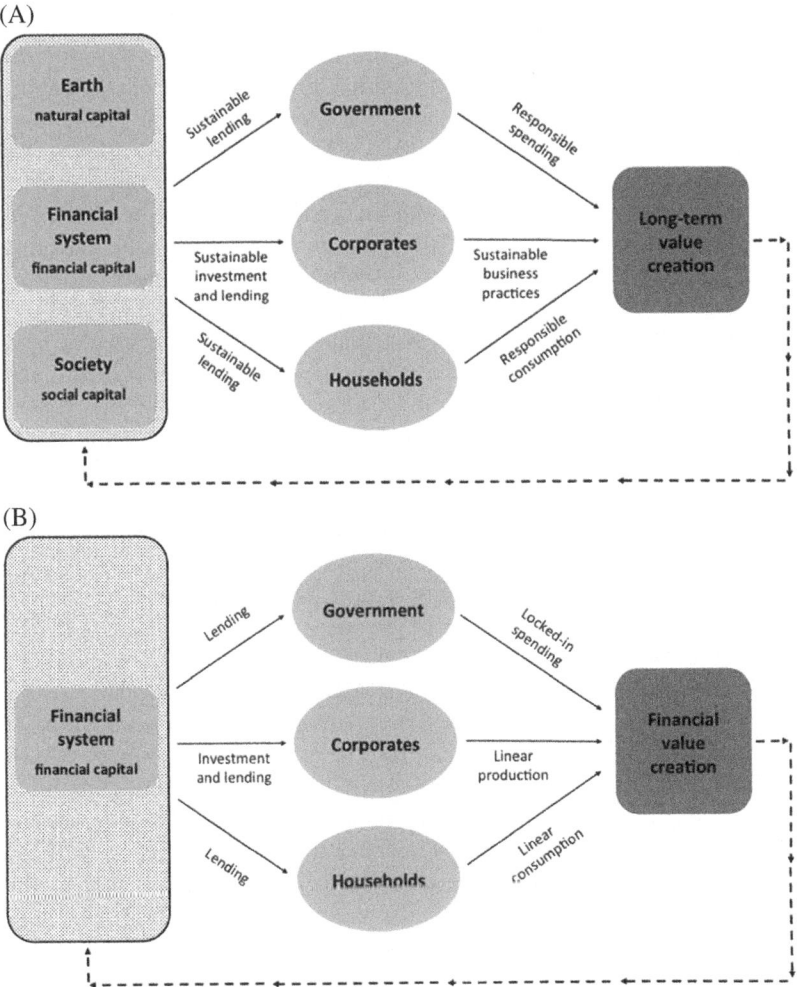

Fig. 1.3 Value creation in traditional finance (Panel A) and sustainable finance (Panel B) (*Source* Schoenmaker e Schramade [2019], op. cit., p. 29)

Within this framework, Sustainable and Responsible Investments (SRI) have emerged as a central component of sustainable finance. SRIs incorporate two fundamental objectives: financial returns that ensure long-term shareholder value and non-financial outcomes that address broader stakeholder interests, such as community well-being and environmental preservation. Increasingly, empirical research supports the notion that SRIs can yield comparable, if not superior, financial returns, challenging the traditional view that sustainable practices compromise profitability (Friede et al., 2015; Khan et al., 2016). This approach recognizes that financial and non-financial value creation are not mutually exclusive, with SRI practices proving viable for both financial and societal benefits. As a result, the SRI market has expanded significantly in recent years, reflecting investors' growing commitment to sustainability-focused portfolios that balance profit with purpose.

To support this trend, various sustainable investment strategies have been developed, including ESG integration, impact investing, and thematic investments. Recognized as global standards by the Global Sustainable Investment Alliance (GSIA), these strategies demonstrate diverse methods for embedding sustainability into investment decisions. Regardless of the approach, each strategy aligns with the core objective of creating long-term value for both investors and society by integrating ESG criteria into portfolio construction and management (GSIA, 2022).

The data in Fig. 1.4 highlights regional disparities in asset growth from 2014 to 2022, with notable trends emerging. Europe and the United States show steady but modest growth (CAGR of 4% and 3% respectively), while Canada (15%) and Australia & New Zealand (30%) demonstrate stronger expansion, likely driven by smaller initial bases and favorable policies. Japan stands out with an exceptional CAGR of 122%, reflecting extraordinary growth from a low starting point in 2014. Growth trends vary across periods: Europe and the U.S. experienced modest early gains but saw declines during 2018–2020, with the U.S. facing a sharp contraction (-51%) in 2020–2022, likely due to the pandemic. Conversely, Australia & New Zealand maintained consistent growth across all periods, while Japan showed sustained, albeit decelerating, increases. These figures underscore the resilience of emerging markets like Australia & New Zealand and Japan, the stability of Europe and the U.S., and the challenges of post-pandemic recovery, particularly in North America.

						GROWTH PER PERIOD				COMPOUND ANNUAL GROWTH RATE (CAGR) 2014-2020
	2014	2016	2018	2020	2022	2014-2016	2016-2018	2018-2020	2020-2022	
Europe (EUR)	€9,885	€11,045	€12,306	€10,730	€12,401	12%	11%	-13%	31%	4%
United States (USD)	$6,572	$8,723	$11,995	$17,081	$8,400	33%	38%	42%	-51%	3%
Canada (CAD)	$1,011	$1,505	$2,132	$3,166	$3,014	49%	42%	48%	-5%	15%
Australia & New Zealand (AUS)	$203	$707	$1,033	$1,295	$1,680	248%	46%	25%	30%	30%
Japan (JPY)	¥840	¥57,056	¥231,952	¥310,039	¥493,598	6692%	307%	34%	59%	122%

NOTE: Asset values are expressed in billions. All figures are in regional currencies. New Zealand assets were converted to Australian dollars.

Fig. 1.4 Growth of sustainable investing assets by region in local currency, 2014–2022 (*Source* GSIA [2022], *Global Sustainable Investment Review 2022*)

Regarding the changes observed during the years under analysis, the report highlights that sustainable investment assets are growing globally, with the exception of Europe, which saw a 13% decrease in the growth of sustainable investment assets during the 2018-2020 period. However, this decline is attributed to significant changes in how sustainable investments are defined under EU legislation, making it difficult to directly compare with other regions of the world and previous versions of the GSIA report. In Canada, the largest increase was recorded for the 2018-2020 period, with sustainable assets growing by over 48%. The United States follows closely behind, with a 42% growth over the same two-year period. Japan comes next, registering a 34% increase from 2018 to 2020. As for Australasia, the report indicates that sustainable assets continued to grow in this region, but at a slower pace compared to the 2016-2018 period, with a 25% growth from 2018 to 2020 compared to 46% growth from 2016 to 2018.

In conclusion, the main findings of the 2020 report show that the United States and Europe continue to represent more than 80% of global sustainable investment assets during the 2018-2020 period. Over the same two years, the shares of global sustainable investment assets remained relatively stable in Canada (7%), Japan (8%), and Australasia (3%). This situation is depicted in Figure 1.5.

Another interesting finding from the 2020 Report is that sustainable investment assets under management (over 35 trillion dollars) represent 35.9% of the total professionally managed assets in the regions covered by this report.

Ethical Banks and Their Comparison with Conventional Banks

The collapse of Lehman Brothers in 2008 marked a critical turning point in the global financial crisis, which severely impacted major developed economies. This crisis was largely attributed to irresponsible lending practices by banks, resulting in an unchecked accumulation of toxic assets. At the time, banks were widely regarded as safe institutions operating under strict regulation, making the severity of the crisis largely unforeseen. In response, governments introduced extraordinary measures, such as financial bailouts and nationalizations, to protect savings and restore stability. Alongside these efforts, the introduction of Basel III regulation represented a significant global response, aimed at strengthening the resilience of the banking sector by improving capital adequacy, risk

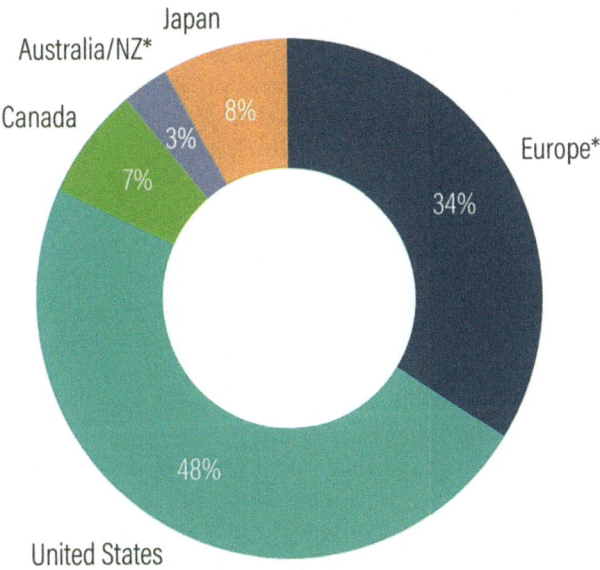

* Europe and Australasia have enacted significant changes in the way sustainable investment is defined in these regions, so direct comparisons between regions and with previous versions of this report are not easily made.

Fig. 1.5 Percentage of sustainable investments by region (*Source* GSIA [2020], *Global Sustainable Investment Review 2020*, p. 10)

management, and liquidity standards. Between 2012 and 2019, additional stringent regulations were implemented to mitigate banking risks further. However, this crisis remains uniquely complex as banks did not technically violate any laws, highlighting deep ethical shortcomings within the financial system.

The central issue, then, was not solely economic but also ethical, involving failures on individual, management, and societal levels. As a result, simply implementing stricter regulations may not be enough to prevent future crises. Instead, the financial sector needs a paradigm shift toward an ethical and sustainable model, an approach that some "ethical banks" have already adopted. These institutions prioritize long-term sustainability and only invest in activities that meet ethical, social, and environmental criteria. Through transparent and democratic governance, ethical banks offer a stark contrast to traditional banking by aligning profit with broader societal and environmental goals.

Ethical banking is not a novel concept. Banks with a social mission have existed for centuries, reflecting the prevailing religious and ethical norms of their times. However, in the last few decades of the twentieth century, a distinct movement of ethical banks emerged, including institutions like South Shore Bank in the United States (1973), Triodos Bank in the Netherlands (1980), and Banca Popolare Etica in Italy (1995). Following the 2008 financial crisis, these banks gained attention and credibility by demonstrating a people-centered, rather than profit-centered, approach to finance.

Several factors contributed to the rise of ethical banks. First, the erosion of public trust in conventional banks, which had engaged in speculative and sometimes unethical practices, played a significant role. Many of these traditional banks relied on government bailouts, causing widespread economic repercussions. This crisis of trust created a demand for banking practices rooted in transparency and responsibility. Second, as awareness of global social and environmental issues has increased, there is growing support for sustainable development within the banking sector, a shift supported by governments, businesses, and financial institutions alike.

Ethical banks are distinguished by their focus on both financial and non-financial impacts, appealing to investors with sustainability concerns. Beyond economic returns, they consider the social and environmental costs of their operations. This twofold impact is essential, as banks' activities—both their direct operations and their allocation of capital—affect society and the environment. For example, the energy consumption associated with running bank offices and data centers is significant, as is the waste produced by day-to-day operations. More importantly, through their financing choices, banks influence the allocation of resources and the overall direction of economic development.

The growing impact of ethical banks suggests their potential to reshape the broader financial landscape. However, academic research on ethical banking remains relatively underdeveloped, likely because this sector still represents a small portion of the global financial system. Most scholars define ethical banking as a commitment to offering products and services that consider environmental and social impacts. Key characteristics include transparency, social responsibility, and environmental stewardship, forming a triad of "integrity, responsibility, and affinity" (Cowton,

2002; San-Jose et al., 2011). Integrity ensures financial inclusion, responsibility demands accountability for social and economic impacts, and affinity fosters close, transparent relationships between depositors and borrowers. To advance understanding of this important topic, future research could employ qualitative methodologies, such as interviews, surveys, and case studies, to explore in greater depth the practices, challenges, and impacts of ethical banking.

Compared to conventional banks, ethical banks differ significantly in transparency and asset allocation. They lend to projects that create social value, often with reduced interest rates, and avoid speculative transactions. By financing socially and environmentally beneficial projects, they support initiatives within the real economy and offer services to clients who might be excluded by traditional banks. Ethical banks are generally smaller, more specialized, and often operate online, which allows them to evaluate project risks more carefully and maintain lower default rates.

Moreover, while conventional banks emphasize profit maximization, ethical banks consider the "triple bottom line" of profit, people, and the planet. They focus on sustainable value creation rather than short-term gains, setting an example for the industry. Though ethical banks currently occupy a niche within the financial sector, they represent a valuable model for sustainable finance. In the event of future financial crises, ethical banks could serve as a blueprint for a more responsible, community-oriented approach to banking, underscoring the need for a financial system that aligns economic activity with environmental and social well-being. This model of ethical banking not only offers a viable alternative but also presents an essential framework for a future-focused, sustainability-driven approach to finance.

The Concept of Corporate Social Responsibility (CSR) and Its Evolution

Corporate Social Responsibility (CSR) is a concept intimately connected to sustainability, reputation, and the role of companies in society. As the mindset around environmental and social issues has evolved, so too has the expectation for businesses to align their values with sustainable practices. Consumers, in particular, are now making purchasing decisions based not solely on products but on the values and social commitments of the companies producing them. This trend is supported by research highlighting how businesses that communicate their CSR strategies effectively and transparently can significantly enhance their corporate reputation and attract consumer loyalty (McWilliams & Siegel, 2001; Porter & Kramer, 2006). Given this evolution, it has become essential for companies to incorporate CSR as a core strategy and manage their image actively to address public expectations of responsibility and ethical conduct.

In management literature, CSR and sustainability are viewed as interdependent. Sustainable development becomes feasible only when organizations are socially responsible, often resulting in improved corporate performance (Carroll & Shabana, 2010). This mutual reinforcement creates a cycle where responsible actions lead to sustainable growth, which, in turn, enhances a company's public image. Consequently, CSR has evolved into a strategic tool through which companies not only fulfill ethical duties but also establish themselves as leaders in social and environmental stewardship. By fostering a strong reputation for responsibility, businesses gain a competitive edge, attracting investors, customers, and talented employees who value alignment with sustainable values (Eccles et al., 2014).

The concept of CSR has been discussed in management literature for decades, beginning in the 1950s with foundational work by economist Howard R. Bowen, often regarded as the "father of CSR." Bowen's seminal book Social Responsibilities of the Businessman (1953) argued that corporations, as social institutions, have obligations beyond profit maximization. Bowen posited that businesses must consider societal welfare, emphasizing voluntary social contributions as part of a broader duty to society. He highlighted that, because businesses operate within society and benefit from its resources and institutions, they bear a responsibility to contribute positively in return. This idea laid the groundwork for CSR to be seen as an integral component of strategic management and not just an ancillary responsibility (Bowen, 1953).

Throughout the 1960s and 1970s, CSR gained traction as an academic field, with scholars such as Keith Davis (1960) arguing that social considerations should take precedence over pure profit motives. William C. Frederick (1960) further asserted that businesses should work toward societal well-being rather than focusing solely on private gains. Joseph McGuire (1963) expanded this notion, arguing that companies have responsibilities beyond economic and legal obligations, reflecting broader social expectations. By the 1970s, academic discourse around CSR had shifted, reflecting the increasing societal concerns about environmental degradation and inequality. During this period, the Committee for Economic Development (CED) issued its influential report, Social Responsibilities of Business Corporations (1971), which called for a redefined corporate mission. The CED proposed that businesses should support "human values" by contributing to social progress, even at the expense of short-term profits. This marked a shift in thinking, where companies were expected to address broader societal issues and support public welfare (CED, 1971).

The CED report also introduced the idea of CSR as a multi-layered concept with three concentric circles: an inner circle of core economic responsibilities, an intermediate circle encompassing broader societal expectations, and an outer circle addressing emerging social issues. This framework, illustrated in Fig. 1.6, emphasized that businesses must adapt to public expectations and engage with stakeholders to create societal value. According to the CED, corporate success hinged on fulfilling these expectations, as public consent underpins the legitimacy of businesses.

In the 1980s and 1990s, CSR became a mainstream topic in management and business ethics literature, with scholars examining how social and environmental responsibilities could be integrated into corporate strategy. The growing awareness of environmental issues led companies to adopt practices aimed at reducing their ecological footprint, further cementing CSR as a strategic element in organizational policy (Carroll, 1979; Sethi, 1975). The emphasis on environmental sustainability paralleled the development of CSR, leading to the establishment of frameworks such as Elkington's Triple Bottom Line (1997), which proposed that companies should measure success not just in terms of economic performance but also environmental and social impact.

In the twenty-first century, CSR has expanded to encompass sustainability, with the concept of ESG (Environmental, Social, Governance) gaining prominence. Studies have shown that ESG performance can enhance a company's long-term financial performance by reducing risks associated with environmental regulations, social unrest, and governance scandals (Friede et al., 2015). The integration of ESG criteria into CSR demonstrates a more comprehensive approach, addressing the complex challenges of modern society and affirming CSR's relevance in the era of global sustainability concerns. According to Soppe (2009), CSR is essential in achieving sustainable development, as it promotes responsible business practices that align with societal and environmental goals.

In conclusion, CSR has evolved from a voluntary initiative aimed at enhancing corporate image to a fundamental component of modern

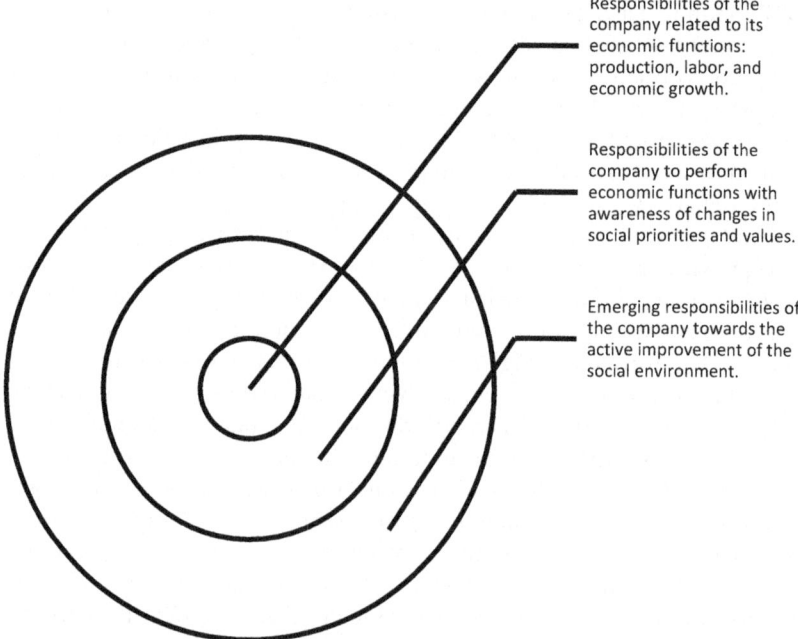

Responsibilities of the company related to its economic functions: production, labor, and economic growth.

Responsibilities of the company to perform economic functions with awareness of changes in social priorities and values.

Emerging responsibilities of the company towards the active improvement of the social environment.

Fig. 1.6 Three concentric circles model of responsibility (*Source* Own elaboration on CED [1971], op. cit., p. 15)

business strategy. The continuous expansion of CSR reflects a growing recognition that businesses must balance profit generation with social responsibility and environmental stewardship to maintain legitimacy and competitive advantage. As companies navigate this shift, they face increasing pressure from stakeholders to implement CSR and ESG practices, which not only satisfy regulatory requirements but also resonate with a public that values transparency and ethical conduct.

According to the CED report, the innermost circle includes the basic and well-defined responsibilities related to efficiently carrying out the company's economic functions, such as production, labor, and economic growth. The middle circle encompasses the responsibility of performing those economic functions with an awareness of the shifting social priorities and values within society. Some examples of this level of responsibility include: environmental protection; working conditions and relationships with employees; stricter customer expectations regarding information, fair treatment, and protection from harm. Finally, the outermost circle represents emerging and still undefined responsibilities that companies should assume to be more actively involved in improving the social environment, such as addressing issues like economic and cultural underdevelopment, urban degradation, and related challenges. The CED's construction of CSR provides a significant contribution as it reveals the perspective of professionals and business leaders on the changing social contract between businesses and society, as well as on the new social responsibilities companies must take on.

As CSR continued to evolve in the 1970s, the literature expanded significantly. One of the most important contributions to this body of work came from Archie B. Carroll in 1979, with his A Three-Dimensional Conceptual Model of Corporate Social Performance. Carroll proposed a four-part definition of CSR, comprising economic, legal, ethical, and discretionary responsibilities. To fully address the range of obligations businesses have towards society, the definition of social responsibility must incorporate all four of these categories. Firstly, companies have fundamental economic responsibilities, which include producing goods and services demanded by society, selling them, and generating a profit. Additionally, companies are expected to operate within the bounds of laws and regulations. In other words, society expects companies to fulfill their economic mission while adhering to legal constraints, thereby developing legal responsibilities. However, while these first two categories incorporate ethical norms, there are many other behaviors expected by society that are not necessarily codified in law. Since society holds expectations beyond legal requirements, it can be said that businesses also have ethical responsibilities. Finally, discretionary responsibilities are those left to individual judgment and voluntary choice. The decision to adopt these responsibilities comes solely from the company's desire to engage in non-mandatory social rules that are not required by law or expected based on ethical principles. Each of these responsibilities constitutes a part of a company's total social responsibility.

Building on these developments, during the 1980s, scholars focused more on research related to CSR and on differentiating publications into alternative concepts and themes, such as Corporate Social Responsiveness, Corporate Social Performance (CSP), public policy, business ethics, and stakeholder theory. It would be a mistake to think that interest in CSR faded during the 1980s. In fact, the key issues surrounding CSR began to be reformulated into alternative theories, concepts, themes, and models. One notable contribution during this decade is an article by Thomas M. Jones, titled Corporate Social Responsibility Revisited, Redefined (1980). Unlike previous scholars, Jones emphasized CSR as a process. Starting from the premise that reaching consensus on what constitutes 'socially responsible behavior' is highly complex, Jones argued that CSR should be viewed not as a set of outcomes but rather as a process. This new perspective formed what the author termed a "revised or redefined" concept. Jones' contribution was valuable and innovative, though it did not end the debate over the content and scope of CSR expected of businesses.

In 1983, in the article Corporate Social Responsibility: Will Industry Respond to Cutbacks in Social Program Funding?, Carroll revisited and refined his four-part definition of CSR, repositioning the discretionary component to include volunteerism and/or philanthropy. The reason for this revision was that the best examples of discretionary activities observed were coming from voluntary and philanthropic initiatives. Also, during the 1980s, research began to show increasing interest in how CSR related to corporate financial performance. Scholars started exploring whether socially responsible businesses were also profitable. If true, this factor would provide further justification for the CSR movement. Some of the empirical studies during this time that attempted to verify the relationship between CSR and profitability include those by Philip Cochran and Robert Wood (1984), as well as Aupperle, Carroll, and Hatfield (1985).

In general, the 1990s saw fewer unique and original contributions to the definition of CSR. During this period, CSR served primarily as a starting point for other related concepts and themes, with most writings aligning with the definitions established in prior decades. The focus of the 1990s centered on research into CSP (Corporate Social Performance), stakeholder theory, and business ethics, each of which developed its own extensive literature. Nonetheless, an interesting study by Carroll in 1991 further revised his four-part CSR model, first developed in 1979. It was through this work that Carroll introduced the now-famous "Pyramid of Corporate Social Responsibility," where economic responsibilities form the foundation upon which all other responsibilities rest. The pyramid builds upwards through legal, ethical, and, finally, philanthropic responsibilities (Fig. 1.7).

The CSR pyramid serves as a metaphor to visually represent the four components of corporate responsibility. The foundation is built on the premise that economic performance underpins everything else. At the

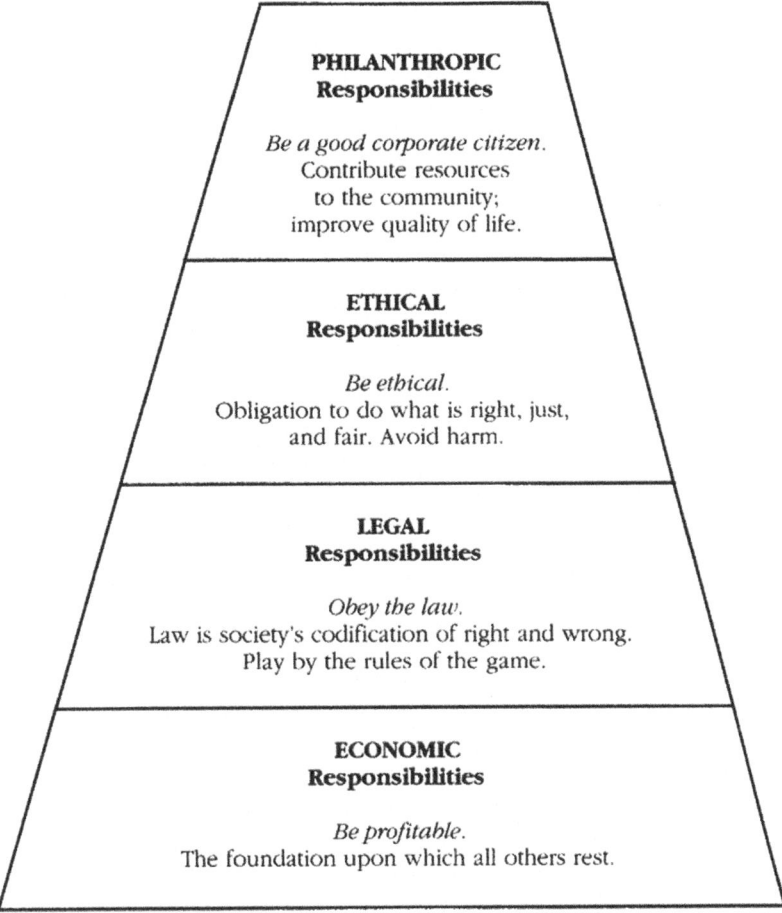

PHILANTHROPIC
Responsibilities

Be a good corporate citizen.
Contribute resources
to the community;
improve quality of life.

ETHICAL
Responsibilities

Be ethical.
Obligation to do what is right, just,
and fair. Avoid harm.

LEGAL
Responsibilities

Obey the law.
Law is society's codification of right and wrong.
Play by the rules of the game.

ECONOMIC
Responsibilities

Be profitable.
The foundation upon which all others rest.

Fig. 1.7 CSR pyramid (*Source* Carroll [1991], op. cit., p. 42)

same time, businesses are expected to comply with the law, as the law codifies what society deems acceptable and unacceptable behavior. In addition, companies have a responsibility to act ethically, meaning they are obligated to do what is right, fair, and just while minimizing harm to stakeholders (employees, consumers, the environment, and others). Lastly, a company must act as a good corporate citizen. This is captured by philanthropic responsibility, where businesses are expected to contribute financial and human resources to the community and improve the quality of life. Carroll (1991) also clarified that companies should not fulfill these responsibilities in a sequential manner, but rather all of them should be met at all times. More specifically, the author stated:

> Stated in more pragmatic and managerial terms, the CSR firm should strive to make a profit, obey the law, be ethical, and be a good corporate citizen.

In the years following Carroll's work, these four categories or domains of CSR, which underpin the pyramid, have been widely adopted by various theorists (Swanson, 1995, Wartick & Cochran, 1985; Wood, 1991) and empirical researchers (Aupperle, 1984, Aupperle et al., 1985, Burton & Hegarty, 1999, Clarkson, 1995, Smith et al., 2001).

Additionally, during the 1990s, the concept of accountability began to take root, emphasizing the need for formal reporting and transparent communication of a company's commitments and achievements in the context of social responsibility towards all stakeholders. As a result, companies were called upon to fulfill a dual fundamental commitment: first, to guarantee an adequate return to shareholders, and second, to pursue social goals by transparently reporting on and communicating their activities. This period saw the emergence of new concepts connected to CSR, enriching managerial vocabulary.

The turn of the millennium brought heightened attention to the environmental impact of many human activities. Actors across the globe—including governments, communities, businesses, organizations, and individuals—have increasingly embraced sustainability principles. In fact, the current era can be summarized by a new term introduced by Paul Crutzen and Eugene Stoermer: Anthropocene. Officially proposed in 2000, the term Anthropocene defines the current geological epoch, in which the Earth's environment, in all its physical, chemical, and biological dimensions, is profoundly shaped by human activity. The term highlights the severe impacts of human actions on global ecosystems and the climate system, underscoring the human capacity to drastically influence geological processes and cause territorial, structural, and climatic shifts.

A key event demonstrating humanity's impact on the global ecosystem is the 2007–2009 financial crisis, discussed earlier. Following this global crisis, standards of responsibility, ethics, and sustainability regained prominence as essential metrics for evaluating corporate behavior. As noted by Carroll (2016), Corporate Social Responsibility has a rich history of literature and academic contributions. Looking to the future, the author remains optimistic, stating:

> The future of CSR, whether it be viewed in the four-part definitional construct, the Pyramid of CSR, or in some other format or nomenclature such as Corporate Citizenship, Sustainability, Stakeholder Management, Business Ethics, Creating Shared Value, Conscious Capitalism, or some other socially conscious semantics, seems to be on a sustainable and optimistic future.

As such, it is expected that in the years to come, there will be a gradual increase in the adoption of CSR practices by companies worldwide. CSR as a management strategy has now become a formalized, common practice integrated into corporate policies and organizational procedures. CSR is recognized as a beneficial approach for both businesses and society. Moreover, in recent years, there has been a significant proliferation of academic studies and research on the subject across various disciplines, and this trend is predicted to continue growing.

Nevertheless, throughout the development of CSR, there have been dissenting opinions. One notable critique came from German economist Theodore Levitt, who raised concerns in his 1958 article "The Dangers of Social Responsibility." Levitt is known for initiating the debate on CSR by criticizing this emerging philosophy. He argued that CSR posed a threat to the foundations of the free market, potentially overshadowing the core functions of businesses and governments. In other words, he believed that assigning social roles to businesses was risky, as their sole purpose should be to generate high levels of profit. Several years later, the distinguished American economist Milton Friedman endorsed a similar position. In his famous 1970 article titled "The Social Responsibility of Business is to Increase its Profits," published in The New York Times Magazine, Friedman echoed Levitt's views, describing CSR as a "subversive" theory of corporate responsibility. According to Friedman, profitability was the highest form of social responsibility, as long as it was pursued ethically and in accordance with the law. Thus, while a company's sole responsibility was to increase profitability and create value for shareholders, the responsibility for addressing social and environmental concerns should fall to the state through legislation. Nonetheless, it is evident that Friedman's perspective, focused on maximizing shareholder wealth, was heavily criticized by CSR proponents. Below is Table 1.1, which presents some of

Table 1.1   CSR definition, academic literature

Author	Definition
Carroll (1979)	The social responsibility of business encompasses the economic, legal, ethical, and discretionary expectations that society has of organisations at a given point in time
Jones (1980)	Corporate social responsibility is the notion that corporations have an obligation to constituent groups in society other than stockholders and beyond that prescribed by law and union contract
Wood (1991)	The basic idea of corporate social responsibility is that business and society are interwoven rather than distinct entities
Baker (2003)	CSR is about how companies manage the business processes to produce an overall positive impact on society
Dahlsrud (2008)	CSR is frequently defined as a concept whereby companies integrate social and environmental concerns in their business operations and interactions with stakeholders on a voluntary basis
Carroll and Shabana (2010)	CSR encompasses strategies that firms adopt to address the social and environmental impacts of their business operations in alignment with their stakeholders' expectations
Aguinis and Glavas (2012)	CSR refers to organizational actions and policies that take into account stakeholders' expectations and the triple bottom line of economic, social, and environmental performance
Sarkar and Searcy (2016)	CSR definitions are highly diverse, reflecting its 'chameleon-like' nature, but commonly emphasize stakeholder engagement, ethical practices, and contributions to sustainable development
Elkington (2018)	CSR, now often reframed as 'sustainability,' integrates social, environmental, and economic dimensions to ensure long-term value creation for society and business alike

Source Own Elaboration

the most representative academic definitions of CSR from the past fifty years.

SUSTAINABLE FINANCE AND ECONOMIC THEORIES

Having delved into the evolution of sustainability and Corporate Social Responsibility (CSR)—outlining their milestones and growing importance in today's economy—it is equally crucial to examine the theoretical frameworks that support this paradigm shift toward sustainability. These theories form the foundation for understanding how and why companies are increasingly expected to adopt socially and environmentally responsible practices. In essence, they provide the "roots" from which modern CSR and sustainability initiatives have grown. The economic theories presented here are central to grasping how social responsibility and sustainability have become integral to corporate strategy and governance. This approach, which draws from the management and economic literature, is essential for interpreting the CSR and sustainability landscape, particularly given that these concepts have varied definitions and are often contextualized differently across disciplines.

As CSR has evolved, it has prompted extensive academic debate, especially around whether companies should integrate social responsibility into their strategies to address stakeholder expectations. This debate has led to a spectrum of theoretical positions, both supporting and opposing the idea that businesses should go beyond profit maximization. These positions often blur the lines between literature on business and society, as corporate purpose and social impact are increasingly intertwined.

Three influential theories that have shaped the discourse on CSR and sustainability are shareholder theory, stakeholder theory, and shared value theory. Each offers unique perspectives on the role of business in society, particularly concerning how value is created and distributed among different stakeholders. Shareholder theory, popularized by Milton Friedman (1970), argues that a company's sole responsibility is to maximize shareholder wealth. This perspective contends that engaging in CSR beyond legal obligations diverts from the primary goal of profit maximization, effectively reducing shareholder value.

Conversely, stakeholder theory, developed by R. Edward Freeman in the 1980s, posits that companies have a duty to address the interests of all stakeholders, not just shareholders. Stakeholders include employees, customers, suppliers, communities, and any other entities affected by corporate actions. According to Freeman, businesses should strive to create value for these groups as well, as they play integral roles in a company's success. Stakeholder theory has been widely embraced in CSR literature, reflecting the growing belief that businesses must balance profitability with social responsibility. This theory is especially relevant in today's economy, where consumer expectations and regulatory pressures emphasize transparency and ethical business practices.

Shared value theory, introduced by Michael Porter and Mark Kramer in 2011, goes a step further by suggesting that companies can achieve economic success by addressing societal challenges. Unlike CSR, which is often seen as separate from core business objectives, shared value theory integrates social impact into a company's competitive strategy, proposing that solving societal issues can drive profitability. By aligning corporate success with positive social outcomes, shared value theory offers a framework for sustainable business practices that generate both economic and societal benefits. This approach emphasizes that companies can unlock new markets, enhance productivity, and strengthen their competitive advantage by addressing social needs, such as environmental sustainability, health, and economic development.

Together, these theories provide a spectrum of approaches to the social role of business. The core question, therefore, becomes not only what companies are responsible for but also to whom they are accountable. In other words, is a company accountable solely to its shareholders, as Friedman posits, or does its responsibility extend to all stakeholders and society as a whole, as suggested by Freeman and Porter? This ongoing debate between shareholder-centric and stakeholder-inclusive perspectives reflects broader societal shifts toward sustainability and ethical business practices, demonstrating how theoretical foundations continue to influence CSR evolution and implementation.

Moreover, these theories offer insight into the future trajectory of CSR. As global challenges related to sustainability intensify, the balance between shareholder and stakeholder interests will likely continue to evolve. For instance, environmental, social, and governance (ESG) considerations are becoming essential factors in investment decisions, pushing companies to adopt broader responsibility frameworks. This alignment of business interests with sustainable practices underscores the relevance of these theories in an era increasingly defined by the pursuit of shared prosperity and environmental stewardship.

Shareholder Theory

Milton Friedman, a Nobel laureate and one of the most influential economists of the twentieth century, is widely recognized for developing Shareholder Theory, a paradigm emphasizing that a corporation's primary responsibility is to maximize profits for its shareholders while operating within the bounds of legal and ethical norms (Friedman, 1970). According to Friedman, a company's social role should be restricted to economic value creation; activities that detract from this goal—such as environmental or social initiatives—are viewed as inefficiencies that reduce shareholder value. This theory posits that companies should maximize capital returns, aligning their actions with shareholder interests and contributing to society by maximizing economic output (Jensen, 2002).

Under this neoclassical framework, any diversion of resources towards non-economic objectives, such as social responsibility or environmental stewardship, represents a misallocation of corporate assets and is deemed unacceptable.

Friedman's critique of Corporate Social Responsibility (CSR) argues that such initiatives impose additional costs that can undermine profitability, viewing CSR as a threat to the economic efficiency that drives market competitiveness (Baden & Harwood, 2013). This approach holds that costs associated with social and environmental activities detract from a firm's primary obligation to enhance shareholder wealth, thus weakening its economic performance (Friedman, 1970). His seminal essay, The Social Responsibility of Business is to Increase Its Profits, published in The New York Times Magazine (1970), remains a cornerstone text, frequently revisited and debated in both academic and business contexts. This work notably encapsulates his viewpoint that business leaders should "engage in activities designed to increase its profits so long as it stays within the rules of the game," suggesting that adherence to legal and ethical standards is sufficient social responsibility (Friedman, 1970).

Friedman's approach aligns with Agency Theory, which outlines a principal-agent relationship where managers, as agents, are tasked with acting in the best interests of shareholders, the principals. Under this model, managers are expected to prioritize profit maximization over social aims (Eisenhardt, 1989). Friedman, however, was not entirely opposed to social goals; he argued that shareholders, managers, and other stakeholders are free to engage in philanthropy and social activities but should not do so with corporate funds. This delineation suggests that the responsibility for addressing societal needs lies with individuals rather than corporations. Indeed, he believed free-market capitalism itself contributes to social welfare by meeting needs and facilitating economic growth (Friedman, 1962).

This perspective, though influential, has been criticized for its limitations, particularly its narrow focus on financial gain at the expense of broader social responsibilities. Scholars such as Edward Freeman have argued for a broader corporate accountability framework, highlighting the need to address the interests of a wider range of stakeholders—employees, customers, communities, and others affected by corporate activities (Freeman, 1984). This broader view challenges Friedman's assertion that shareholder value alone drives social welfare, instead proposing that businesses should actively contribute to societal well-being

(Donaldson & Preston, 1995). Furthermore, the 2008 financial crisis highlighted the potential dangers of a profit-centric corporate model, underscoring the risks associated with neglecting ethical considerations in favor of short-term profits (Crotty, 2009).

The critique of CSR was also shaped by Friedman's political context. As an advocate for free-market capitalism, he perceived socialism as a fundamental threat to economic freedom and prosperity. His views were part of an ideological movement to safeguard capitalism from socialist ideals that advocated for state intervention in economic affairs (Donna, 2021). Additionally, Friedman cautioned against the risk of managerial overreach, where corporate leaders might pursue personal interests under the guise of social responsibility, thus eroding shareholder control (Jensen & Meckling, 1976). He also warned against monopolistic tendencies, advocating for regulatory oversight to maintain competitive markets.

Friedman's shareholder-focused perspective has become a fundamental principle in business education and economic discourse. However, the global shift towards sustainability has led to increasing criticism of this approach. Scholars now argue that the focus on shareholder value alone is insufficient to address contemporary challenges, such as environmental degradation and social inequality (Schoenmaker & Schramade, 2019). Modern theories suggest that shareholder value creation and environmental, social, and governance (ESG) objectives are not mutually exclusive but can be pursued simultaneously to achieve sustainable value creation (Clark et al., 2015). For instance, Creating Shared Value theory posits that companies can drive profitability by addressing societal needs, bridging the gap between shareholder interests and societal well-being (Porter & Kramer, 2011).

In today's context, businesses are increasingly pressured to integrate sustainability into their operations. Regulatory developments, such as the EU's Sustainable Finance Disclosure Regulation, require companies to disclose their ESG impacts, reflecting a shift in expectations that challenges Friedman's focus on profit maximization as the sole corporate objective (European Union, 2019). This shift indicates a growing recognition of the role of sustainable practices in risk management and value creation, suggesting that profit-seeking and social responsibility can coexist within a responsible and sustainable business framework (Khan et al., 2016).

In conclusion, while Friedman's Shareholder Theory has profoundly influenced corporate governance and economic thought, evolving societal expectations now advocate for a more holistic corporate purpose. This new paradigm acknowledges that companies can—and should—create value not only for shareholders but also for a broader set of stakeholders. Rather than abandoning profit as a goal, the modern perspective integrates it with ethical and sustainable business practices, viewing shareholder value and social responsibility as complementary, rather than conflicting, objectives (Eccles et al., 2020). This evolution highlights the necessity for businesses to adapt their strategies to address environmental and social challenges, ultimately contributing to a more resilient, inclusive, and sustainable economic system.

Stakeholder Theory

Stakeholder Theory directly challenges the notion that a corporation exists solely to maximize shareholder profits. R. Edward Freeman, in his influential article "Stockholders and Stakeholders: A New Perspective on Corporate Governance" (1983), introduced this idea as an alternative framework, proposing that businesses have responsibilities not just to shareholders but to a broad range of stakeholders who are affected by, or can affect, corporate activities (Freeman & Reed, 1983). Freeman's theory posits that the objective of a corporation should be to create value for all stakeholders, including shareholders, employees, customers, suppliers, communities, and even natural resources, given their critical role in sustaining business operations and societal welfare (Donaldson & Preston, 1995).

Freeman further articulated this approach in his seminal book, Strategic Management: A Stakeholder Approach (1984), emphasizing that companies should prioritize the needs of stakeholders to achieve long-term success and ethical legitimacy. This paradigm was innovative as it reframed the corporate world's understanding of responsibility, advocating that the survival and growth of an organization depend on harmonious, mutually beneficial relationships with all stakeholders (Jones, 1995). Rather than perceiving stakeholders as external to a corporation's objectives, Stakeholder Theory embeds them within the organization's core strategy and governance, acknowledging that companies operate within a "system of dynamic relationships" where every stakeholder has an impact on the organization (Freeman et al., 2007).

The theory categorizes stakeholders into primary and secondary groups. Primary stakeholders, or "narrow stakeholders," are those on whom the company's survival directly depends—such as employees, key customers, suppliers, shareholders, and regulators (Freeman, 1983). Secondary stakeholders, or "broad stakeholders," include those who influence or are influenced by the company's operations but are not essential to its immediate survival, such as public interest groups, unions, and community representatives (Clarkson, 1995). By organizing stakeholders in this way, Freeman highlighted the complexity of stakeholder relationships and the necessity of active management to ensure organizational resilience and sustainability.

Stakeholder Theory has grown in influence, evolving from a fringe idea to a widely accepted approach that integrates ethical, social, and economic dimensions into corporate governance. Jones and Wicks (1999) describe four foundational assertions of the theory, namely that: (1) businesses have relationships with multiple constituencies (stakeholders), (2) outcomes of these relationships affect both the company and the stakeholders, (3) the interests of all legitimate stakeholders are valuable, and (4) the theory emphasizes that managerial decisions should consider the interests of all stakeholders. This approach broadens corporate accountability and balances traditional profit motives with the need to address the expectations of various stakeholders, redefining corporate success as a balance between financial performance and social impact (Clarkson, 1995; Donaldson & Preston, 1995).

A notable linkage between Corporate Social Responsibility (CSR) and Stakeholder Theory is provided by Carroll (1991), who argued that CSR could be more comprehensively defined through a stakeholder lens, giving specificity to the often vague concept of "social" responsibilities in business. By identifying specific groups to whom companies are accountable, the stakeholder concept effectively "gives a name and face" to those who benefit from corporate CSR efforts, embedding CSR within the strategic planning and ethical obligations of business operations (Carroll, 1991; Wood, 1991). This alignment between CSR and Stakeholder Theory reinforces the idea that businesses not only impact society but have ethical responsibilities toward it, forming a foundation for CSR practices and stakeholder management frameworks today (Carroll & Buchholtz, 2008).

In analyzing Stakeholder Theory, Donaldson and Preston (1995) identified four central perspectives: descriptive, instrumental, normative, and managerial. Descriptively, the theory reflects the reality of corporate structures where multiple stakeholders are interdependent. Instrumentally,

Stakeholder Theory serves as a means to improve corporate performance, suggesting that a stakeholder-centered approach can enhance profitability. Normatively, the theory provides ethical guidelines, positing that companies have inherent obligations to stakeholders that go beyond financial metrics. Lastly, the managerial perspective combines these elements, encouraging management to adopt strategic interventions that balance stakeholder needs to achieve corporate objectives. This framework allows businesses to address stakeholder expectations pragmatically while reinforcing the moral legitimacy of corporate actions, especially in contexts where ethical considerations are critical to long-term success (Jones et al., 2002).

The normative aspect of Stakeholder Theory is particularly significant, as it asserts that stakeholders possess legitimate claims on the corporation and that their interests hold intrinsic value, independent of their ability to affect shareholder returns. This interpretation offers a moral foundation for corporate governance, where ethical principles guide managerial actions to respect stakeholder rights and address their interests fairly. By integrating these ethical standards, companies can foster trust and long-term support from their stakeholders, which enhances social capital and sustains competitive advantage in an increasingly interconnected world (Harrison & Freeman, 1999).

Stakeholder Theory thus embodies a paradigm shift in corporate governance, where organizations prioritize the welfare of all their stakeholders, not just their shareholders. The theory aligns with the concept of managerial capitalism, which envisions a business environment where ethical considerations are integral to corporate strategy. By redefining the corporate objective as one of shared value creation rather than pure profit maximization, Stakeholder Theory lays the groundwork for sustainable, ethical business practices that can adapt to complex social and environmental demands (Porter & Kramer, 2011). As global economies face unprecedented challenges, such as climate change and economic inequality, Stakeholder Theory's integrated approach to management offers a valuable framework for achieving business goals while fostering social progress.

In conclusion, Stakeholder Theory provides a comprehensive framework that balances the pursuit of profit with the ethical responsibility to address the needs of all individuals and groups connected to the company. As businesses continue to confront societal expectations for sustainable and responsible behavior, this theory remains relevant, guiding organizations toward a future where economic performance and social value creation are inextricably linked.

Shared Value

The discussion in this section reflects a paradigm shift from the general idea of corporate responsibility towards society at large to a more focused perspective on meeting the specific demands of various interest groups, or stakeholders, involved in or affected by a company's activities. This shift is radical, as it reframes the purpose of corporate value creation from a generalized commitment to the "common good" to one that prioritizes stakeholders' needs based on their power, legitimacy, and urgency (Mitchell et al., 1997). This approach introduces a nuanced model of accountability and recognizes that a company's long-term success depends not only on shareholders but also on balancing diverse, and sometimes competing, stakeholder interests.

Closely related to this paradigm shift is the emergence of Corporate Shared Value (CSV), a concept introduced by Porter and Kramer as an evolution beyond traditional Corporate Social Responsibility (CSR). Their seminal 2011 article, Creating Shared Value: How to Reinvent Capitalism—and Unleash a Wave of Innovation and Growth, presents CSV as a new framework for addressing the growing tension between business and society. Unlike CSR, which often remains peripheral to a company's core strategy and is frequently focused on reputation, CSV embeds social and environmental objectives within the company's core business model, thus creating both economic and social value (Porter & Kramer, 2011). Porter and Kramer argue that CSR, though beneficial, tends to treat social and environmental issues as secondary considerations. CSV, in contrast, redefines profitability by linking it to societal benefits, aiming to make companies catalysts of social progress while generating long-term economic value.

The Corporate Shared Value model proposes three strategies through which companies can achieve both economic and societal benefits: (1) reconceiving products and markets to meet underserved needs or address social issues; (2) redefining productivity in the value chain by increasing resource efficiency, improving working conditions, and enhancing local supply chains; and (3) supporting local clusters through knowledge sharing and development to improve the economic and social environment in which they operate (Porter & Kramer, 2011). For example, companies can focus on sustainable resource management by reducing waste and emissions, thereby lowering production costs while also contributing to environmental sustainability. This shift to a CSV approach is particularly relevant today as consumers and investors increasingly prioritize companies that engage in sustainable and responsible practices (Clark et al., 2015; Eccles et al., 2014).

A fundamental aspect of CSV is its alignment with capitalism's primary goals—creating economic value while contributing to societal well-being. Porter and Kramer argue that businesses have historically confined value creation to short-term financial gains, which neglects broader societal needs and risks long-term sustainability. By reconciling business and societal interests, CSV creates a new form of capitalism that promotes "the right kind of profits," those that advance rather than undermine societal well-being (Porter & Kramer, 2011). This approach underscores a shift in capitalism towards what has been called "sustainable capitalism," which not only emphasizes profit but also advocates for ethical and sustainable practices (Benn & Bolton, 2011; Elkington, 1997).

This new model of creating shared value has significant implications for business strategy, governance, and performance measurement. Unlike CSR, where societal engagement is often detached from the core business, CSV integrates societal impact as a vital component of corporate strategy. This strategic integration enables companies to innovate by developing products, processes, and practices that are sustainable and directly aligned with their primary business objectives (Lee, 2008). Companies like Nestlé and Unilever have adopted CSV principles to drive growth in emerging markets, enhance local supply chains, and improve environmental sustainability, demonstrating how CSV can provide a competitive advantage in the global marketplace (Lash & Wellington, 2007; Porter et al., 2012).

Porter and Kramer's framework positions CSV not as a replacement for CSR but as a progression toward a deeper integration of social value within business objectives. The focus is not solely on corporate philanthropy or compliance-driven CSR practices but on creating economic value through initiatives that simultaneously benefit society. This distinction is crucial, as it clarifies that the purpose of CSV is not altruistic; instead, it leverages social needs as drivers of economic success. By fostering shared value, companies contribute to social progress in ways that are financially sustainable, potentially restoring public trust in business and establishing a foundation for resilient and inclusive economic growth (Pfitzer et al., 2013).

Another essential component of CSV is the recognition of long-term benefits and sustainability in value creation. CSV proposes that addressing social and environmental challenges is not just a moral imperative but a business opportunity that can generate enduring competitive advantage. For instance, efforts to reduce energy use or improve labor conditions can enhance operational efficiency and brand reputation, thereby attracting ethically conscious consumers and investors (Eccles & Serafeim, 2013; Kanter, 2011). In this sense, CSV represents a departure from the "profit-first" mentality, replacing it with a holistic, integrative approach that sees profitability and social progress as mutually reinforcing.

The implementation of CSV also requires a comprehensive rethinking of corporate governance, as companies must ensure that value creation benefits all stakeholders, including shareholders, employees, and communities. Donaldson and Preston's (1995) model of Stakeholder Theory aligns with CSV, asserting that corporate strategies should account for all legitimate stakeholders, not just shareholders. By emphasizing the value of each stakeholder, CSV enriches corporate purpose, thereby fostering trust, loyalty, and long-term engagement (Clarkson, 1995). For businesses, this shift underscores the importance of aligning operations with societal expectations and ethical standards to build resilient relationships across diverse stakeholder groups (Harrison & Freeman, 1999).

In conclusion, while CSV may not be a solution to all societal problems, it offers a robust framework for aligning business practices with societal needs in a way that is both profitable and sustainable. By incorporating societal impact into the core business model, CSV enables companies to address the structural weaknesses in traditional capitalism that have often led to environmental degradation, social inequality, and public distrust. As more companies adopt the CSV framework, it has the potential to redefine capitalism itself, fostering a more equitable, sustainable, and prosperous global economy where business success is measured not only by financial gain but also by the positive impact on society and the environment (Porter & Kramer, 2011).

The Origin, Meaning, and Function of ESG Factors

After discussing the concepts of sustainability, sustainable development, the new approach to sustainable finance, socially responsible investments (SRI), corporate social responsibility (CSR), and the academic theories that have heightened awareness of environmental, social, and governance issues, it is time to delve into the central topic of this work: the three pillars of sustainability, also known as ESG factors.

The term "ESG" was officially coined in 2004 with the publication of the renowned "Who Cares Wins" report, initiated by the United Nations Global Compact. The aim of this report was to group together the three main pillars of ethical finance, encompassing various themes and specific evaluation objectives. The "Who Cares Wins" report emerged from a conference organized by then UN Secretary-General Kofi Annan (who served from 1997 to 2006) with over fifty CEOs of leading financial institutions worldwide. This initiative aligned with the earlier Global Compact and was supported by the International Finance Corporation and the Swiss government. Following the conference, the report "Who Cares Wins: Connecting Financial Markets to a Changing World" was published by eighteen financial institutions with the goal of "developing guidelines and recommendations for better integrating environmental, social, and corporate governance [ESG] issues into asset management, brokerage services, and associated research functions." In short, the goal was to integrate ESG considerations into the financial sector, particularly in capital markets.

The report's signatories emphasized the crucial belief that companies paying close attention to these issues could increase shareholder value by properly managing risks, anticipating regulatory interventions, and accessing new markets. Moreover, this focus on ESG factors could have a significant impact on reputation and brand value, key components of corporate value. Thus, the report aimed to raise awareness among financial market players about the need to initiate a broader discussion on these topics, fostering mutual understanding, active collaboration, and constructive dialogue.

The report's significant ESG recommendations led supporting institutions to commit to achieving the following broad goals:

- Stronger and more resilient financial markets;
- Contribution to sustainable development;

- Greater awareness and mutual understanding among stakeholders;

Improved trust in financial institutions (UN, 2004).

The concept of ESG builds on the broader tradition of sustainable and responsible investment (SRI), which has existed for much longer. However, unlike SRI, which is based on ethical and moral criteria and often employs negative screening (e.g., avoiding investments in companies involved in alcohol, tobacco, or firearms), ESG investments are founded on the premise that ESG factors hold financial relevance. In fact, ESG factors cover a wide range of issues that are traditionally outside financial analysis but can have a significant financial impact. As will be discussed further, this includes issues such as how companies respond to climate change, how effectively they manage water resources, their commitment to employee health and safety, protection against accidents, or their treatment of suppliers.

Notably, around the same time as the "Who Cares Wins" report, the UNEP Finance Initiative (UNEP FI) released the so-called "Freshfield Report," which demonstrated that ESG issues are relevant for financial evaluation. Over the following years, numerous academic studies confirmed that strong corporate sustainability performance is associated with positive financial outcomes. For example, works by George Serafeim, Bob Eccles, and Ioannis Ioannou—which will be discussed later in this study—highlight this connection. Today, one of the main streams of academic literature is focused on the relevance of ESG factors for corporate financial performance and profitability. Whether in developed or emerging economies, most studies agree that commitments and disclosures related to ESG have positive impacts on companies' financial performance. This, in turn, has contributed to growing awareness of the importance of ESG information for evaluating risks, strategies, and operational performance.

Now, let us explore the meaning behind the acronym "ESG." This term refers to three distinct concepts, each representing a different domain of social responsibility: "E" for Environmental, "S" for Social, and "G" for Governance. ESG has become a standard for defining a sustainable investment approach. Essentially, ESG denotes a framework in which investments are assessed based on environmental, social, and corporate governance criteria. However, the concept of ESG lacks a single, universally accepted definition and is used in various contexts.

Typically, it is applied in corporate procedures to identify a set of relevant elements—environmental, social, and governance factors—that help measure the long-term sustainability of investments, integrating them with traditional economic and financial parameters. Conversely, ignoring ESG factors poses the risk of overlooking a range of opportunities and risks for the company and its stakeholders, which ESG is designed to capture effectively. In short, ESG, if managed innovatively and responsibly, aims to become an indispensable asset of the modern enterprise. Indeed, environmental, social, and governance responsibilities operate as a set of dynamic capabilities that constitute a new competitive factor for businesses.

Furthermore, it is worth noting that the academic community often uses CSR (corporate social responsibility), ESG, and ESGEE (economic, governance, social, ethical, and environmental sustainability) interchangeably, as—as noted by Basen, A. et al. (2008)—there is no single definition for ESG. One of the most frequently cited frameworks for defining ESG factors is the United Nations Principles for Responsible Investment (UN PRI), which has been referenced in many reports, including directives and regulations from the European Parliament and the European Commission. The UN PRI provides guidelines on defining the three ESG domains and the issues to be included in each, encouraging financial institutions to integrate these factors into their decision-making processes. In these principles, environmental, social, and governance factors are linked to three distinct, yet interconnected, areas of "social awareness."

In other words, ESG factors represent the three pillars of sustainability—environmental, social, and economic—that are introduced into the financial sector. Today, the acronym ESG is becoming increasingly familiar even outside strictly financial circles, reaching broader segments of society. The three dimensions encapsulated by ESG (environmental, social, and governance) play a crucial role in verifying, measuring, monitoring, and supporting a company or organization's sustainability commitments, particularly in relation to product purchasing or investment decisions. ESG criteria are used to measure and monitor companies' performance regarding their environmental impact, social responsibility, and the quality of their internal governance practices. Specifically, ESG refers to a set of standards and criteria for measuring an organization's environmental, social, and governance activities. These criteria represent the operational standards upon which a company's activities and processes should be based to achieve specific sustainability objectives—environmental, social, and governance-related. Their fundamental importance lies in the fact that investors use these criteria to evaluate and make informed investment decisions.

The major benefit of ESG factors is that they allow for precise measurement of environmental, social, and governance performance through standardized, recognized parameters, ensuring objective evaluations and comparability of results.

To summarize, ESG criteria are employed in the financial world to evaluate companies' environmental, social, and governance performance, as well as the sustainability of investments. As Aguinis (2011) emphasized, ESG can be considered an evolution of the concept of CSR (corporate social responsibility) because it specifies three fundamental types of relationships between stakeholders and the company: environmental, social, and governance. In this sense, many scholars argue that ESG can be seen as the modern "idea" of social responsibility.

The growth and increasing prevalence of ESG-driven investments can also be seen as a sign that markets and society, in general, are undergoing a green transition. Concepts of evaluation are evolving in tandem with this shift. By focusing on ESG factors, companies are moving away from the typical industrial-era models, where pollution was not alarming, labor was merely a cost factor, and the dominant strategy was to maximize short-term profits. The new corporate landscape, however, promotes smarter, healthier, and more environmentally friendly products and services. From an investor's perspective, ESG data is crucial for identifying the most virtuous and stable companies in the long term, avoiding those at risk of underperforming or failing. ESG investments allow individuals to express their values and see their savings and investments directly reflected in their preferences. For policymakers and governments, ESG commitments represent an opportunity for market-driven development where the 'common good' is not lost in the race for short-term profit.

Since the introduction of the ESG concept in 2004, these issues have become increasingly prominent worldwide, and this trend seems to be steadily growing. In Europe, ESG considerations have long been a priority, but now they are gaining significant traction in Asia as well. Moreover, a survey conducted by Klynveld Peat Marwick Goerdeler (KPMG, Survey of Corporate Sustainability Reporting, 2017) revealed that many countries with the highest levels of corporate responsibility reporting—where ESG disclosure is considered integral—are located in the Asia–Pacific region (Asia and Oceania). The same survey indicated that the overall social responsibility reporting rate in the Asia–Pacific region was 78%, compared to the global reporting rate of 52% (versus 52% in the Middle East and Africa). Furthermore, the survey also showed that an increasing number of Asian investors are incorporating ESG factors into their analyses and investment decisions, as this can attract foreign investors and thus increase returns.

In general, what can be observed is the growing awareness among the public and companies about environmental, social, and governance issues. From 2004 to the present, the number of companies adopting sustainability strategies and disclosing their ESG information globally has significantly increased. This practice can bring great opportunities for companies, such as gaining a competitive advantage, improving operational efficiency, enhancing reputation, reducing waste, and improving shared value and sustainability performance. As a result, ESG-based investments limit exposure to a wide range of risks and increase portfolio resilience.

Additionally, it has been proven that ESG criteria are becoming increasingly popular among investors, not only for the tangible benefits in terms of performance but also for the reputational advantages they offer. As such, the financial world is increasingly incorporating these factors into investment decisions. However, each of the three dimensions of sustainability—environmental, social, and governance—has its own evolutionary path. Recently, BlackRock's 2020 study "Sustainability Goes Mainstream" reveals that, among the three criteria (environmental, social, and governance), the environmental dimension is the most prominent. Indeed, 425 investors involved in the study (active in 27 countries and representing $25 trillion) indicated that the importance of the "E" (environmental) criterion is expected to grow from 88 to 89% over the next 3–5 years. During the same period, the social criterion is expected to grow from 52 to 58%. Finally, the governance criterion is projected to decline in focus, from 60 to 53%.

In the next section, we will delve into the individual ESG criteria.

*The E Dimension (*Environmental*)*

The "Environmental" criteria, represented by the letter "E" in ESG, pertain to a company's environmental commitments and the impact it has on the environment in which it operates. In other words, this pillar focuses on issues such as climate change, CO2 emissions, deforestation, air and water pollution, land use, and biodiversity loss. Essentially, this criterion evaluates a company's or organization's performance in areas like energy efficiency, greenhouse gas emissions, waste management, water usage, and resource management. The Environmental criterion measures the risks and opportunities that businesses face in relation to the aforementioned environmental aspects.

Of all the ESG factors, the environmental criterion likely garners the most attention, as the serious environmental challenges we face today are widely visible. It only takes a moment's reflection for each of us to recognize, at least optimistically, the visible consequences of the environmental issues that afflict our planet daily. Additionally, the media tends to focus heavily on environmental issues. Movements such as Fridays for Future, founded in 2018 by Greta Thunberg, illustrate the growing public attention to environmental protection.

The European Banking Authority (EBA, 2020) defines environmental risks as the financial risks arising from banks' exposure to counterparties that may either contribute to or be impacted by climate change or other forms of environmental degradation (such as air and water pollution, freshwater scarcity, land contamination, loss of biodiversity, and deforestation). Similarly, climate-related risks refer to financial risks stemming from banks' exposure to counterparties potentially involved in or affected by climate change. The EBA (2020) further notes that there is some overlap between environmental and climate risks, with environmental risks encompassing climate risks. For instance, while climate change can lead to environmental degradation, not all environmental degradation results from climate change.

It is interesting to note that environmental risks manifest through three main transmission channels: physical, transition, and liability transmission channels (EBA, 2020):

Physical risks pertain to the tangible effects of climate change or environmental factors, which are categorized into: a) acute physical effects, resulting from specific events such as extreme weather conditions (storms, floods, wildfires), or heatwaves that can damage production facilities and disrupt supply chains; and b) chronic physical effects, which arise from longer-term trends like temperature changes, sea-level rise, reduced water availability, and biodiversity loss.

Transition risks concern the shift towards a low-carbon, climate-resilient, or environmentally sustainable economy. These risks stem from government policies, technological changes, and stakeholder expectations, which may force businesses or industries to alter their business models, leading to asset devaluation, higher management costs, and reduced profit margins.

Liability risks refer to risks faced by banks due to exposure to counterparties that could be held accountable for the negative environmental, social, or governance impacts of their activities. Essentially, liability risk pertains to claims for compensation by individuals or businesses for losses incurred due to ESG factors.

Both physical and transition risks have a direct impact on the value of tangible and intangible assets, reducing their worth. Additionally, they result in higher costs and lower revenues, significantly influencing a company's financial stability and, consequently, its creditworthiness.

Given that this thesis focuses on the impact of ESG factors on bank credit and examines the relationship between banks and businesses, these are the areas that will be further explored. Financial institutions are increasingly aware of the physical and transition risks related to climate change, to which their portfolios are exposed. Numerous studies indicate that climate-related losses could be substantial:

> In the case of large publicly listed banks in the Eurozone, Battiston et al. (2017) estimate that climate risk could lead to total losses amounting to around 30% of the banks' capital.

The European Commission (March 8, 2018, Action Plan: Financing Sustainable Growth) warns that at least half of Eurozone banks' assets are exposed to risks related to climate change.

On a global scale, Dietz et al. (2016) suggest that the value at risk of financial assets, due to the projected rise in global temperatures by the end of the century, could reach $2.5 trillion.

Regarding corporate environmental performance and access to credit, Capasso et al. (2020) conducted research examining the relationship between CO_2 emissions and the 'distance to default' of 458 companies from 2007 to 2017. Their findings reveal that higher CO_2 emissions per unit of product are associated with a greater risk of corporate failure.

Moreover, concerning the relationship between the environmental factor and the cost of debt, several authors have analyzed this using data related to green bonds. For instance, Gianfrate and Peri (2019) found that green bonds have lower spreads compared to traditional bonds of the same type, based on a sample of European corporate bonds issued between 2013 and 2017. This suggests that the sustainable projects underpinning green bonds are perceived as less risky by investors. Another interesting study by Eichholtz et al. (2019) examined a sample of real estate investment trusts (REITs) and found that commercial mortgages backed by environmentally certified properties have lower interest rates than those without such certifications.

From this brief analysis, it is evident that the environmental factor plays a critical and dominant role in assessing the sustainability commitments of companies and organizations. Essentially, adopting policies that do not adequately address environmental impacts can lead to significant challenges for a company, including negative repercussions for its access to bank financing.

The S Dimension (Social)

The "Social" criteria, represented by the letter "S" in ESG, pertain to the social impact of a company and measure its relationship with key aspects such as the local community, employees, suppliers, customers, and the broader communities it engages with. In essence, this social dimension includes issues such as gender policies, human rights protection, labor standards, product safety, health and safety policies to prevent accidents, relationships with civil communities, public health, and income distribution. All these elements influence employee satisfaction and overall corporate reputation. Therefore, the social factor essentially relates to a company's internal relationships with employees and its external interactions with stakeholders such as communities, unions, customers, and suppliers. Like environmental factors, social issues can represent both risks and opportunities, making their measurement essential for evaluating companies.

Social risks are generally more difficult to identify compared to environmental risks. According to the European Banking Authority (EBA, 2020), social risks arise from a bank's exposure to counterparties that may be negatively impacted by social factors. Consequently, social factors relate to issues such as rights, welfare, and the interests of individuals and communities, all of which can influence the activities of a bank's counterparties. Today, these concerns are receiving increasing attention in corporate strategies and the operational activities of companies, banks, and their counterparties.

The social factor has been the subject of extensive academic research, with many empirical studies highlighting the relationship between social factors and debt capital. Specifically, most studies focus on corporate social responsibility (CSR). For example, Goss and Roberts (2011), using a sample of 3,996 loans to U.S. firms, found that companies with social responsibility issues face higher borrowing costs (between 7 and 18 basis points more) compared to companies with stronger social responsibility practices. This finding aligns with another significant study by Cooper and Uzun (2015), which examined a large sample of U.S. companies across various sectors between 2006 and 2013. They documented that firms with strong CSR initiatives have lower debt financing costs. This relationship holds across all sectors, though it is particularly pronounced in the manufacturing and financial industries. Furthermore, they found that managerial ownership weakens the benefits of strong CSR practices on debt costs, suggesting that companies may benefit from strengthening their CSR activities to reduce borrowing costs, particularly when managerial ownership is less significant.

Another major area of research focuses on the effect of CSR on corporate credit ratings. Attig et al. (2013) found that credit rating agencies tend to assign higher ratings to companies with strong social performance. Their findings suggest that CSR performance conveys important non-financial information, which rating agencies use to assess a firm's creditworthiness. Additionally, this research highlights that investments in CSR can lead to lower financing costs, derived from higher credit ratings. Similarly, Dorfleitner et al. (2020) found varying impacts of CSR on credit ratings depending on whether companies are based in North America, Europe, or Asia. While the extent of the impact differs, the risk-reduction effect of CSR is observed across all regions. The authors noted that social factors have a greater influence on credit ratings in North America and Europe, while the effect is less pronounced for Asian companies.

Fatemi et al. (2015) also found that, under certain conditions, CSR expenditures can create value for a firm, affirming that CSR investments pay off by creating value. More broadly, many studies suggest that CSR, through its ability to build a strong corporate image and reputation, effectively enhances a firm's performance. For example, Sun et al. (2014) empirically examined the relationship between CSR and an important financial indicator: default risk. Their results confirmed that CSR has a strong effect in reducing a company's risk of default. According to the authors, socially beneficial activities, while entailing costs, result in overall gains through lower debt costs and higher credit ratings. This finding aligns with the perspective of Porter and Kramer (2002), who argue that "in the long run... social and economic goals are not inherently conflicting but closely intertwined." In other words, if economic activities drive social change, corporate social responsibility often yields economic gains. This idea, commonly referred to as "doing well by doing good," is illustrated by Sun et al. (2014) in the context of corporate default risk.

Overall, the studies discussed above indicate that enhancing a company's involvement, commitment, and transparency in social factors is advantageous, as it generates benefits in terms of better access to credit and lower debt capital costs.

The G Dimension (Governance)

The final criterion, represented by the "G" in ESG, refers to "Governance," which involves corporate management guided by ethical principles and fair practices. The governance factor encompasses aspects such as board independence and composition, respect for shareholders' rights, executive compensation policies, control procedures, anti-competitive practices, transparency in decision-making, protection of minority shareholders, and compliance with the law. In essence, this third dimension of ESG relates to corporate governance practices, which can significantly impact the performance and stability of a company.

Similar to social factors, identifying the impact of governance factors and risks is not always straightforward. According to the European Banking Authority (EBA, 2020), governance risks stem from a bank's exposure to counterparties that may be negatively affected by governance factors. These factors refer to the governance choices made by counterparties, including how they integrate ESG factors into their policies and practices. Governance risks can manifest in various ways. For example, an inadequate code of conduct or a lack of anti-money laundering actions within a company can impair its ability to generate positive returns or secure financial resources, potentially triggering reputational risks. If such governance issues become public, they could result in a loss of trust from customers and investors, legal expenses, sanctions, and long-term damage to the company's business operations.

Governance also plays a crucial role in ensuring that environmental and social considerations are integrated into corporate strategy. Recognizing the potential impact of climate change and environmental risks—such as physical, transitional, or liability risks—is seen as an indicator of good governance. Conversely, neglecting these factors in strategic planning can create additional governance risks. Therefore, banks must recognize the impacts and risks posed by these issues. On the corporate side, companies are responsible for incorporating environmental and social considerations into their governance, fostering a corporate culture that supports equality, inclusion, fair labor standards, and community engagement. By doing so, companies signal strong governance practices.

Regarding academic research on the governance factor, many scholars have focused on the relationship between governance practices and a company's financing costs. For example, Bradley et al. (2011) found that companies implementing policies to enhance executive accountability—such as professional liability insurance—tend to achieve higher credit ratings and lower yield spreads. Such measures incentivize corporate leaders to avoid litigation risks by adopting low-risk strategies, benefiting both the company and investors. Similarly, Zhu (2014) demonstrated that companies with strong corporate governance are associated with lower costs of both equity and debt capital in an international context. However, the relationship between corporate governance and the cost of debt is stronger in countries with weaker legal protection, low transparency, and poor governance quality.

Another strand of research has examined the impact of governance and CSR factors on credit ratings. For instance, a study by Lin et al. (2020) on Taiwanese firms found that companies practicing strong corporate governance and engaging in CSR activities tend to improve their credit ratings. However, the benefits in terms of credit rating improvements are much more pronounced for large firms compared to smaller or family-owned businesses.

While the existing literature on ESG factors is extensive, a particularly interesting study by Derbali et al. (2020) explores whether creditors consider governance characteristics when determining the cost of debt, using a sample of 486 U.S. firms from 1998–2017. Their findings suggest that audit quality and financial expertise serve as key informational tools for creditors, as they provide insights into the reliability and accuracy of financial information. The presence of these elements significantly reduces the cost of debt. Moreover, the study indicates that creditors place a higher value on the presence of independent directors on the board, which leads to a reduction in borrowing costs.

In conclusion, while governance may seem less tangible than environmental or social factors, it plays a pivotal role in determining the financial stability and long-term success of companies. Strong governance, coupled with transparency and ethical practices, not only enhances corporate performance but also lowers the cost of capital and builds trust with investors and stakeholders.

ESG FACTORS FROM THE PERSPECTIVE OF COMPANIES

After understanding the concept of "ESG factors" and delving into the three different dimensions of sustainability, the focus now shifts to examining these factors within the context of businesses. In fact, ESG topics are becoming increasingly relevant for companies, with more and more businesses either meeting these criteria or actively working towards doing so.

Over the past two decades, companies have increasingly strengthened their sustainability efforts with a forward-looking vision, focusing on environmental, social, governance, and financial goals. Corporate sustainability has thus become one of the main research topics in both managerial and financial perspectives. From a managerial research standpoint, many studies highlight that ESG issues (environmental, social, and governance criteria) bring significant benefits to both a company's value and its financial performance. From a financial perspective, most researchers stress the need to integrate ESG objectives into credit scoring evaluations—the automated system used by banks and financial intermediaries to assess loan applications—as well as into lending policies adopted by financial institutions. Evidence shows that a company's ESG commitments can help mitigate credit rating risks in two key ways: ESG factors influence borrowers' cash flows and the estimation of a company's probability of default. Consequently, strong ESG performance should positively impact credit ratings, as high ESG performance leads to better credit ratings. In other words, banks may offer lower rates to companies demonstrating added value and sustainability from their ESG efforts, while companies can access credit at favorable rates and financially support their ESG activities.

Today, environmental, social, and governance objectives play a crucial role for companies that face the imperative of achieving strong performance across all these dimensions, improving corporate sustainability in the process. In the last decade, sustainability reporting has also improved, especially through the Global Reporting Initiative (GRI) guidelines, which promote the disclosure and reporting of ESG performance.

Corporate sustainability management involves an ethical approach to sustainability, focusing on the sustainable use of resources, the preservation and protection of ecosystems, and a human rights-based perspective. Strategic business decisions regarding resource allocation should move from an instrumental approach—where social, human, and environmental interests are simply accounted for—to an intrinsic mindset that creates value for all stakeholders. Thus, sustainability is based on the synergy between social, economic, and environmental improvement, grounded in principles of fairness and ethics. This perspective should be embedded into a company's organizational structure through a collaborative approach that includes all stakeholders. To be sustainable and adopt a long-term perspective, companies should set both financial and ESG goals while aiming to reduce overall business risk. More specifically, environmental, social, governance, and financial goals should be pursued by applying criteria of effectiveness (achieving established commitments) and efficiency (optimal allocation of financial resources). These theoretical concepts are inherently connected, as corporate sustainability directly relates to the company's own sustainability from an ethical, economic, and financial standpoint. As Sharfman and Fernando (2008) explain, "If a company makes a more 'green' (i.e., more efficient) use of its resources by generating less pollution and waste from the resources employed, it will also be more economically efficient."

One of the biggest challenges for a company within the realm of sustainability is establishing a decision-making process for investments that balances environmental, social, governance, and financial aspects, all while maintaining a long-term perspective. So, what should companies do to achieve this complex objective? They should implement investment strategies that respect capital budgeting principles, ensuring at least a financial return on the invested capital while paying close attention to the broader societal impact of their investment decisions.

For clarity, Fig. 1.8 provides a summary of the internal organizational structure of companies where financial, environmental, social, and governance objectives are aligned to go beyond mere profit maximization and improve corporate sustainability by involving the interests of all stakeholders.

This figure provides a clearer view of the framework for sustainability in businesses. On one side, companies integrate environmental, social, and governance (ESG) objectives into their investment processes, which are qualitative factors that cannot immediately be quantified in monetary terms. On the other side, the capital structure, including equity financing and debt financing, allows businesses to raise the funds needed to operate. The question then arises: what differentiates a sustainable business from a traditional one in practice? Naturally, sustainability impacts all three aspects of the ESG framework. Being a sustainable company means, for example, using waste from other processes as input, reducing or eliminating the use of new materials extracted from the earth, creating products that can be reused in other processes, eliminating waste that cannot be recycled or returned to its natural state, minimizing energy use, and relying on energy from renewable sources.

Undoubtedly, achieving these goals is no easy task for a business. However, consumers and society at large increasingly value companies

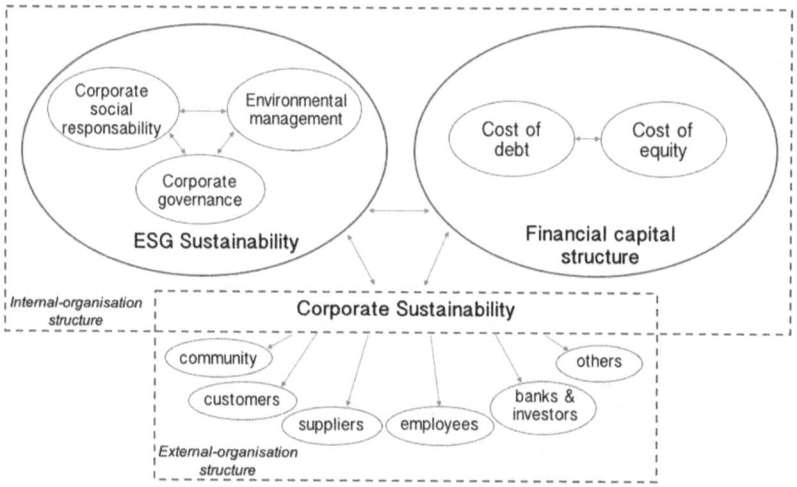

Fig. 1.8 Conceptual framework of the internal structure of sustainable enterprises (*Source* Devalle, A. et al. [2017], op. cit., p. 118)

that adopt sustainable policies and actively engage in ESG. Today, sustainability is recognized as a strategic advantage that can greatly contribute to a company's competitiveness. In this context, we can distinguish between "pioneers," the companies that have embraced visionary and bold sustainable strategies, and "newcomers," which are just beginning their journey towards sustainability. The pioneers—representing less than 10% of all companies—have already begun to reap the benefits of their long-term commitment to sustainability, such as cost reductions, process efficiency, improved reputation, and competitive advantage. Overall, sustainable companies benefit in terms of social sustainability by gaining trust from both internal and external stakeholders and environmental sustainability through green, low-impact decisions that lower costs. In other words, sustainability enables businesses to integrate their strategies, products, and processes with environmental, social, and governance considerations, creating additional value with a long-term perspective.

It is important to emphasize that sustainability is not just about installing solar panels or organizing charity dinners. Often, sustainability efforts are limited to energy-saving practices, waste management, or eco-friendly initiatives, focusing only on the environmental aspect. While this is not wrong, it restricts the transformative potential that a broader approach to sustainability can offer society. True sustainability is not a fad or a fleeting trend—it is a necessary choice for businesses to keep up with increasingly aware consumers and stricter regulatory frameworks. This new business approach is revolutionary, significantly affecting a company's business model, processes, and products. Being sustainable means embracing a corporate philosophy that requires profound transformation across all areas of the business, with a focus on sustainable development. Moreover, sustainability must address all three dimensions—economic, environmental, and social—as they are interdependent. For example, environmental and social sustainability, by enhancing a company's reputation and brand trust, can become a lever that distinguishes the business from its competitors and potentially improves revenues and profit margins.

This is the essence of the "sustainable corporation" model, defined as a company that bases its actions on sustainability-focused mission, philosophy, values, strategies, policies, processes, interactions, relationships, and products. This definition reflects the three "P's" of the Triple Bottom Line model introduced by Elkington in 1997: Profit, Planet, and People. The sustainable business model involves:

A reformulation of the company's business model (the "Profit" aspect), which means pursuing the firm's economic sustainability while developing a new approach to environmental and social issues. The company's mission may also evolve beyond profit to focus on shared well-being.

A revision of products and processes across the entire value chain (the "Planet" aspect), aiming to optimize products and reduce environmental impacts through more efficient technologies. A circular economy approach should guide the entire supply chain.

The adoption of a social purpose (the "People" aspect), integrating the company's mission with a vision centered on the well-being of all stakeholders. This enables the company to create shared value, improving not only its economic performance but also its environmental and social outcomes.

Incorporating sustainability into a company requires a long-term project that involves investment, participation, time, and courage. Otherwise, companies risk falling into the trap of greenwashing. For businesses, sustainability must be a conscious choice—a strategic response to remain competitive in a changing market and among evolving customers. Today, consumers are increasingly aligning with brands that reflect their values, including respect for the environment, sustainable economic and social development, and ethical practices. Therefore, companies must recognize these shifts and adapt their business models towards a new sustainable economy, one that considers not only economic and financial aspects but also the environmental and social impacts of their decisions.

The key point is that this approach shouldn't be adopted merely to ease guilt or for superficial reasons—it must be understood as a real, additional competitive advantage. The change society is experiencing is radical and unprecedented, and companies, given their crucial role in every country's economy, must actively lead this transformation. This evolution must happen here and now, as our future depends on it.

The Role of Banks in the Sustainable Transition

Banks play an essential role in advancing the transition toward a sustainable economy. As key financial intermediaries, they significantly influence capital allocation, steering funds toward environmentally and socially responsible investments and shaping broader market shifts (Frost et al., 2019). By directing financial resources, banks act as critical agents supporting sustainability initiatives and guiding sustainable development (Campiglio, 2016).

A primary method through which banks contribute to this transition is by incorporating ESG (Environmental, Social, and Governance) factors into lending and investment criteria. By prioritizing firms with strong ESG performance, banks not only encourage responsible practices across sectors but also mitigate their own risks associated with unsustainable operations (Jiang & Muradoğlu, 2021). This alignment with ESG principles helps promote efficient resource use and reduces the environmental footprint of their clients (Bauer & Hann, 2010).

Beyond allocating capital, banks play a vital role in assessing and managing ESG-related risks, particularly as climate change and social disparities pose increasing financial threats. Banks now face the task of evaluating the potential impacts of climate-related disasters on asset values and of identifying risks tied to clients that do not adhere to evolving social and governance standards (Battiston et al., 2017). Effectively managing these risks helps banks protect their portfolios and clients, fostering more responsible business practices across the economy.

Additionally, many banks are taking proactive steps by offering green financial products, such as green bonds and sustainability-linked loans, which are designed to finance projects with environmental or social benefits, like renewable energy and community development (Flammer, 2021). Through these innovative products, banks incentivize sustainable business practices, generating positive environmental and social outcomes (Banga, 2019).

The regulatory environment is also a driving force, pushing banks to enhance their sustainability efforts. International regulations, such as the EU Taxonomy for Sustainable Finance and the Corporate Sustainability Reporting Directive (CSRD), mandate greater transparency around ESG risks and establish clear standards for sustainable finance. This regulatory landscape ensures that banks remain accountable for their ESG impacts and strengthens their role in financing the green transition (HLEG, 2018).

In sum, banks are indispensable to the global transition toward a sustainable and equitable economy. Through capital allocation, risk management, green product innovation, and compliance with emerging regulations, they are integral players in promoting ESG practices across industries. Their sustained commitment to these practices is essential for achieving the broader goals of environmental sustainability, social equity, and responsible governance.

CHAPTER 2

The European Regulatory Framework: Policies and Regulations for Sustainable Finance

Abstract This chapter explores the European Union's pivotal role in integrating sustainability into economic and financial strategies. Anchored by international agreements such as the Paris Climate Agreement and the UN's 2030 Agenda, the EU has emerged as a global leader in sustainable finance. Key initiatives include the European Green Deal, which targets climate neutrality by 2050, and the 2018 Action Plan on Financing Sustainable Growth, which aims to reorient capital flows toward sustainable investments. The chapter also discusses the evolving role of financial markets in mitigating climate risks, promoting transparency, and supporting small and medium-sized enterprises (SMEs). The EU's multi-faceted approach emphasizes collaboration, regulatory frameworks, and the development of global standards to advance sustainability and address pressing climate challenges.

Keywords Sustainable finance · European Green Deal · Action plan on sustainable growth · Paris Climate Agreement · Sustainable Development Goals (SDGs) · Financial transparency · Climate risk management

© The Author(s), under exclusive license to Springer Nature Switzerland AG 2025
N. Del Sarto, *ESG Factors and Financial Outcomes in Banks*,
https://doi.org/10.1007/978-3-031-87748-3_2

INTRODUCTION

The urgency of sustainability considerations is increasingly recognized across sectors, with finance at the forefront due to its capacity to allocate resources towards sustainable projects. As global awareness around environmental, social, and governance (ESG) issues intensifies, the European Union (EU) has taken decisive steps to embed sustainability into its economic and financial strategies. Anchored in major global agreements like the Paris Climate Agreement and the United Nations 2030 Agenda, with its 17 Sustainable Development Goals (SDGs), the EU has not only positioned itself as a global leader in sustainability but has also set a model for other countries (European Commission, 2019; Schoenmaker & Schramade, 2019).

One of the EU's most notable efforts in this area is the European Green Deal, an ambitious policy package aimed at making Europe the first climate-neutral continent by 2050. The Green Deal recognizes that to address climate change effectively, economic growth must decouple from resource use, emphasizing a shift towards a circular economy that promotes resource efficiency and waste reduction (European Commission, 2019). Supporting the Green Deal, the EU Action Plan on Sustainable Finance, introduced in 2018, underscores the importance of reorienting capital flows towards sustainable investments and creating a regulatory environment that prioritizes ESG integration across financial markets. This plan represents a turning point, aiming to harmonize the sustainable finance landscape by encouraging transparent ESG disclosures, establishing standards, and fostering a sustainable investment culture across member states (EU Technical Expert Group on Sustainable Finance, 2019).

In light of the significant changes in the global context—ranging from heightened climate risks to the COVID-19 pandemic—the European Commission announced on July 6, 2021, its commitment to intensifying sustainable finance efforts. This announcement came with the intention to solidify the EU's leadership in setting global sustainable finance standards and recognized that, since 2018, the urgency of addressing climate risks through financial mechanisms had increased markedly (de Sousa, 2021). The revised framework proposed six intervention areas, which include expanding legislative support for sustainable finance and enhancing the resilience of the financial system to ESG-related risks. This push for resilience reflects a growing recognition of climate risks as material financial risks, in line with studies linking climate-related events to economic instability and systemic risk in financial systems (Batten et al., 2016; Monasterolo & de Angelis, 2020).

The EU's sustainable finance strategy also emphasizes inclusivity, with a focus on supporting small and medium-sized enterprises (SMEs) and individual consumers. SMEs are vital to the EU economy, and empowering them with sustainable finance tools can facilitate a broader, inclusive transition across the Union. Additionally, the plan seeks to ensure that the financial sector actively contributes to achieving sustainability goals, particularly through the expansion of green financial products and incentives that align with sustainable practices (European Commission, 2021). The EU's ambition to strengthen the financial sector's sustainability role is complemented by a commitment to regulatory integrity, aimed at safeguarding the EU financial system's transparency and reliability in the face of ESG-related risks (Busch, 2020).

As the EU spearheads the establishment of international sustainable finance standards, it seeks to ensure that its partner countries align with these principles, fostering global sustainability efforts. By supporting countries in adopting sustainable finance practices, the EU aims to create a cohesive approach to sustainability that transcends national borders, addressing the inherently global nature of issues like climate change (Schoenmaker, 2017). Through this multi-pronged strategy, the EU's approach to sustainable finance underlines the importance of collaboration, transparency, and robust regulatory frameworks as pillars for a successful transition to a green economy.

The following sections will outline the European Union's legislative progress in sustainable finance, beginning with an in-depth review of the 2018 Action Plan, which is widely regarded as a cornerstone of the EU's sustainable finance strategy. By examining these regulatory developments, this chapter highlights the evolving role of finance in achieving the EU's ambitious sustainability objectives, ultimately positioning finance as a central lever in the shift towards a resilient, sustainable economy.

FROM THE PARIS AGREEMENT TO THE ACTION PLAN ON FINANCING SUSTAINABLE GROWTH

In light of the adoption of the Sustainable Development Goals (SDGs) as part of the United Nations' 2030 Agenda in September 2015, the European Union has embarked on its own path to integrate sustainability into the financial world, signing the Paris Climate Agreement in December of the same year.

The Paris Agreement is the first legally binding international treaty on climate and climate change, concluded at the Paris Climate Conference (COP21) in December 2015, among the member states of the United Nations Framework Convention on Climate Change (UNFCCC). Its primary goal is to limit global warming to well below 2 °C, preferably to 1.5 °C, compared to pre-industrial levels. To achieve this long-term objective, countries aim to reach the global peak of greenhouse gas emissions as soon as possible to achieve a climate-neutral world by the mid-century. The importance of the Paris Agreement in the multilateral climate change process lies in its binding nature, which unites all nations in a common cause to undertake ambitious efforts to combat climate change and adapt to its effects. In other words, for the first time, nearly all the world's nations have come together to jointly combat climate change and its catastrophic consequences. Specifically, the Paris Agreement was adopted by the 197 member states of the UNFCCC on December 12, 2015. Today, according to the United Nations, 191 states (out of 195 signatories), including all EU member states, are officially part of this agreement.

The Paris Agreement was signed by these nations to outline an action plan to counter and limit global warming, climate change, and their severe effects. In this context, participating governments agreed on several key areas, including:

> Climate mitigation and greenhouse gas reduction: The long-term goal (Article 2 of the Paris Agreement) is to keep the global average temperature increase well below 2 °C above pre-industrial levels, and to aim to limit the increase to 1.5 °C to avoid the more frequent occurrence of global warming's impacts. To this end, it is emphasized that efforts should be made to achieve the global peak of emissions as soon as possible (Article 4)—although this may take longer for developing countries—and to rapidly reduce emissions thereafter, using the best available scientific knowledge, to achieve

a balance between greenhouse gas emissions and anthropogenic absorption in the second half of this century. Additionally, countries adhering to the Agreement are required to submit national climate action plans (Article 3), also known as Nationally Determined Contributions (NDCs).

Transparency (Article 13) and the global stocktake (Article 14): Countries commit to meet every five years to assess collective progress and ensure transparency of actions taken, informing the public accordingly. These reports will also contribute to a global stocktake that tracks progress against commitments.

Adaptation (Article 7): Countries must strengthen resilience and reduce vulnerability to ongoing climate change, contributing to sustainable development. Developing countries, in particular, should receive continued international support for adaptation, as they are the most vulnerable and least resourced.

Loss and damage (Article 8): The Paris Agreement acknowledges the importance of preventing, minimizing, and addressing losses and damages associated with the adverse effects of climate change, through mechanisms like risk insurance and early warning systems.

The role of cities, regions, and local authorities: Non-state actors—such as cities, subnational authorities, civil society, the private sector, and others—are called to intensify efforts to meet climate goals, supporting emission reduction actions, and promoting regional and international cooperation.

Support: The EU and other developed countries commit to supporting climate action to reduce emissions and strengthen resilience to climate impacts in developing countries. The collective goal of mobilizing $100 billion per year by 2020, extended to 2025, remains in place, after which a new and higher goal will be set.

It is clear that achieving the Paris Agreement's objectives requires significantly ramping up climate related interventions. Since its entry into force in 2015, numerous low-carbon solutions and greener markets have emerged, driven by the determined efforts of many countries, regions, cities, and companies toward carbon neutrality. These zero-carbon solutions are increasingly adopted, becoming a defining factor for competitiveness across all economic sectors, particularly in key industries like energy and transport.

The EU has always been at the forefront of international climate efforts, both for brokering the Paris Agreement and for continuing to lead in this area. The Agreement was formally ratified by the European Union on October 5, 2016, which led to its entry into force 30 days later, on November 4, 2016. For the Paris Agreement to take effect, at least 55 Parties to the Convention (UNFCCC), representing at least 55% of global greenhouse gas emissions, had to deposit their instruments of ratification, acceptance, approval, or accession with the Depositary. Today, all EU member states have signed and ratified the Agreement, committing to becoming the world's first climate-neutral economy and society by 2050. To achieve this, within the framework of the Paris Agreement, the European Commission has agreed on the need to reach three key objectives by 2030, as set out in EU Regulation 2018/1999:

> A reduction of greenhouse gas emissions by at least 40% compared to 1990 levels, with different targets for each country (Italy's target is a 33% reduction from 2005 national levels);
> An increase in the share of renewable energy consumption to at least 32%;
> An improvement in energy efficiency by at least 32.5% compared to 1990 levels, with specific energy-saving obligations for each member state.

To achieve these goals, the same EU Regulation identifies five key areas or "energy dimensions": energy security; the internal energy market; energy efficiency; decarbonization; and research, innovation, and competitiveness.

In this context, the financial sector's role is crucial to achieving the EU's goals of reducing greenhouse gas emissions and improving energy efficiency as part of the transition to a more sustainable economy. For this reason, European institutions have decided to reform financial markets. Specifically, in December 2016, the European Commission established a High-Level Expert Group on Sustainable Finance (HLEG) tasked with developing guidelines and recommendations to support the growth of sustainable finance in Europe.

A further milestone in EU policy was the report "Financing a Sustainable European Economy," published by HLEG on January 31, 2018, which highlighted two critical needs for the European financial system:

Improving the contribution of finance to sustainable and inclusive growth and mitigating climate change;
Strengthening financial stability by integrating environmental, social, and governance (ESG) factors into investment decisions.

These two imperatives emphasized by HLEG are particularly urgent, given the growing risks associated with climate change, environmental degradation, and other sustainability issues, which can no longer be ignored.

In response to HLEG's recommendations, on March 8, 2018, the European Commission introduced the "Action Plan on Financing Sustainable Growth," a roadmap outlining a series of measures, activities, and specific actions with corresponding deadlines that engage all market actors (investors, managers, and intermediaries). The primary objective of the "Action Plan on Financing Sustainable Growth" is to increase and support investments in sustainable projects by promoting the integration of ESG criteria to develop a financial system that fosters sustainable development. In other words, the Action Plan marks a fundamental step toward sustainability, serving as a means to implement the Paris Climate Agreement and the United Nations 2030 Agenda with its 17 Sustainable Development Goals (SDGs), both adopted in 2015.

Through the Action Plan on Sustainable Finance, the European Commission aims to achieve three fundamental objectives:

Redirecting capital flows towards sustainable investments or economic activities, to foster more sustainable and inclusive growth. Europe currently faces an annual investment gap of nearly €180 billion to meet its 2030 climate and energy targets. Additionally, the European Investment Bank (EIB) estimates that the total annual investment gap in sectors like transport, energy, and resource management infrastructure has reached €270 billion. Therefore, current investments are insufficient to support an environmentally and socially sustainable economy.

Managing financial risks stemming from environmental and social issues, such as climate change, resource depletion, environmental degradation, and social inequalities. Environmental and especially climate-related risks are not yet being adequately considered by the financial sector. With the rise of natural disasters impacting the environment, banks are exposed to greater losses, as companies more affected by climate change or dependent on depleting natural resources generate lower profitability. Between 2007 and 2016, the economic losses caused by extreme weather conditions increased by 86%, amounting to $129 billion in 2016—an alarming figure.

Enhancing transparency and fostering a long-term approach in economic and financial activities, which implies making decisions with long-term goals or consequences. For the financial system to function effectively, market actors must operate transparently. In this regard, corporate transparency on sustainability issues is essential, as it enables market actors to properly assess long-term value creation and corporate sustainability risk management.

The European Commission's Action Plan sets out ten specific actions to be implemented across Europe. Each of the three key objectives listed above corresponds to one or more of these actions. The analysis and in-depth examination of the ten actions contained in the Action Plan will be the focus of the following section (Fig. 2.1).

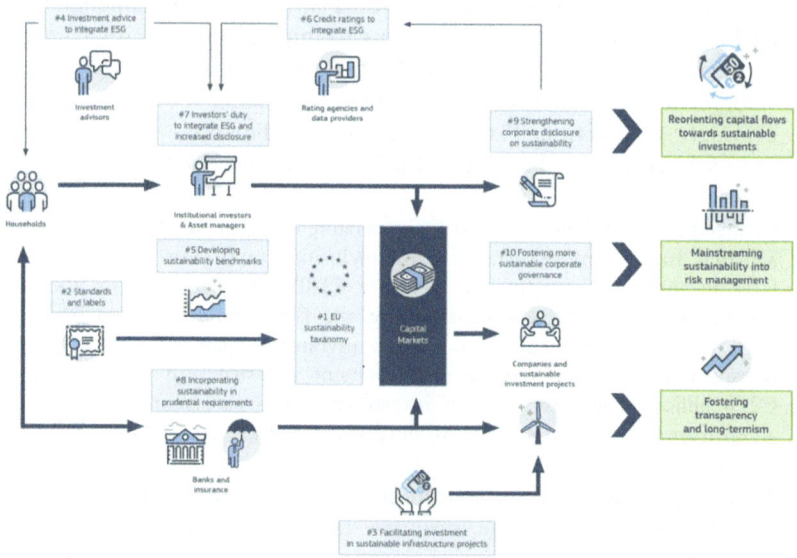

Fig. 2.1 Action Plan on financing Sustainable Growth (*Source* European Commission, "*Action plan on sustainable finance*", p. 21)

THE ACTION PLAN ON FINANCING SUSTAINABLE GROWTH DEL 2018

Building on the first of the three fundamental objectives that the European Commission aims to achieve through the Action Plan on Financing Sustainable Growth—redirecting capital flows towards sustainable investments—five specific actions have been identified.

Action 1: Establish a unified EU classification system for sustainable activities. This action is arguably the most urgent and important of the Action Plan. To direct financial flows toward more sustainable economic activities, the term "sustainable" must be clearly understood, defined, and shared across Europe. The first action seeks to create a unified system, or taxonomy, at the European level to define and classify sustainable economic activities. This "European taxonomy" for sustainable finance will provide financial and non-financial companies, investors, and all economic actors with a common language for sustainability, including the establishment of standards, labels, and reference indices for sustainability. This will create a shared European framework offering detailed information on relevant sectors and activities based on selection criteria, thresholds, and metrics, while also addressing risks associated with greenwashing.

Given the complexity and technical expertise required to define a comprehensive European taxonomy that covers climate, environmental, and social aspects, achieving this goal will take time, as ongoing market, technological, and environmental developments must also be considered. The taxonomy will be developed gradually. A technical expert group on sustainable finance—established by the Commission—will, in an initial phase (by the first quarter of 2019), publish a report outlining a taxonomy for activities related to climate mitigation, adaptation, and certain environmental activities. Subsequently (by the second quarter of 2019), a taxonomy will be developed for the remaining environmental and social activities, taking into account all associated risks and negative impacts.

Action 2: Create standards and labels for sustainable financial products. Through this second action, the European Commission aims to establish standards and quality certifications for sustainable financial products to ensure the credibility and transparency of financial markets while also improving investor confidence. Additionally, EU standards and labels for sustainable financial products would make it easier for investors seeking green products to access them. The Action Plan specifically mentions green bonds, a tool used by operators (companies, banks, organizations, etc.) to borrow from investors to finance or refinance "green" projects or activities. An EU standard for green bonds would facilitate the proper direction of greater investment toward green projects. The Commission's technical expert group on sustainable finance is tasked with preparing a report on a European green bond standard by the second quarter of 2019, followed by the Commission specifying the content of the prospectus for green bond issuances by the same deadline.

Once the sustainability taxonomy is adopted, the Commission will assess the potential use of the EU Ecolabel regulation to create an optional labeling regime for certain financial products at the EU level.

Action 3: Promote investment in sustainable projects. To achieve a more sustainable economic model, private capital must be directed toward sustainable projects. One such area for investment is infrastructure, which, according to an OECD study (Investing in Climate, Investing in Growth, 2017), is responsible for 60% of greenhouse gas emissions. In addition to infrastructure, many other sustainable projects exist, such as those related to improving energy efficiency or expanding renewable energy sources. However, not all EU countries or sectors have the same capacity to implement such projects. Therefore, the European Commission aims to provide more advisory support for promoting environmental, climate, and social projects and to adopt further measures to enhance the efficiency of investment support tools for EU countries. The Commission intends to utilize the European Fund for Strategic Investments (EFSI 2.0, extended until 2020) and the European Investment Advisory Hub.

Furthermore, this action extends beyond EU member states. The EU's External Investment Plan (EIP) has been launched to promote sustainable investments in partner countries such as Africa and regions neighboring Europe.

Action 4: Integrate sustainability into financial advice. Financial intermediaries, such as investment firms and insurance product distributors, play a crucial role in redirecting the financial system toward sustainability. Existing regulations such as MiFID II and IDD already require these intermediaries to offer products suitable to clients' needs. Under the Action Plan, by the second quarter of 2018, the Commission will amend the delegated acts of MiFID II and IDD to ensure that sustainability preferences (including environmental, social, and governance factors) are considered during product selection and suitability assessments. Investment firms and insurance product distributors should inquire about clients' sustainability preferences and take these into account when recommending financial or insurance instruments. The Commission will also invite the European Securities and Markets Authority (ESMA) to include sustainability preferences in its guidelines for assessing product suitability.

Action 5: Develop sustainability benchmarks. Financial benchmarks are used to price financial instruments and other assets, and investors closely monitor them as they provide a way to track and measure performance, helping assess investment activities. Through Action 5, the Commission intends to introduce ESG benchmarks (by the second quarter of 2018) to ensure reliability and transparency, empowering investors to make more informed decisions. The methodologies adopted by benchmark providers in constructing sustainability benchmarks should be more transparent. Additionally, the Commission aims to propose measures to harmonize sustainability benchmarks. The Action Plan also refers to low-carbon benchmarks. The Commission's expert group is tasked with developing a methodology for such benchmarks by the second quarter of 2019.

The second key objective of the Action Plan on Financing Sustainable Growth—integrating sustainability into financial risk management—is associated with three specific actions:

> Action 6: Better integrate sustainability into credit ratings and market research. In recent years, market research companies and credit rating agencies have worked to improve their assessments of companies' environmental, social, and governance (ESG) performance and their risk exposure to sustainability issues. These assessments are crucial because they promote greater transparency between issuers and investors and enable a more sustainable allocation of capital. Credit ratings provide investors with assessments of the creditworthiness of companies and public institutions. Through Action 6, the Commission intends to stimulate the integration of sustainability criteria (ESG factors) by credit rating agencies and market research companies. The Commission will do so in three ways: (1) by evaluating with stakeholders, starting in the second quarter of 2018, whether to require credit rating agencies to explicitly integrate sustainability factors into their assessments; (2) by asking ESMA to encourage credit rating agencies to fully incorporate sustainability and related risks; and (3) by conducting a study on sustainability ratings and market research by the second quarter of 2019.

Action 7: Clarify the obligations of institutional investors and asset managers. EU legislation already requires institutional investors and asset managers to act in the best interest of their final investors or beneficiaries—this is known as the fiduciary duty. Action 7 focuses on introducing sustainability criteria into the definition of fiduciary duty, also requiring the consideration of sustainability risks in investment decisions made for final investors. To implement this action, the Commission will present a legislative proposal by the second quarter of 2018 to explicitly define the obligations of institutional investors and asset managers concerning sustainability: integrating sustainability issues and related risks into the investment decision-making process. The ultimate goal is to increase transparency so that final investors have more information about their investment choices.

Action 8: Integrate sustainability into prudential requirements. This action pertains to the banking and insurance sectors. It is crucial to consider sustainability risks in these sectors, as banks, insurance companies, and pension funds represent the primary source of external financing for the European economy and play a key role in the transition toward a more sustainable economy. The Commission will assess whether to integrate climate-related risks and other sustainability factors into prudential regulation, calibrating banks' capital requirements to ensure that the reliability and effectiveness of the current European prudential framework are not compromised. The goal is to identify capital requirements that better reflect the risks associated with banks' and insurance companies' sustainable activities. Additionally, the Commission will request, in the third quarter of 2018, an opinion from EIOPA (the European Insurance and Occupational Pensions Authority) on the impact of prudential rules on sustainable investments by insurance companies.

The third and final fundamental objective of the Action Plan on Financing Sustainable Growth—promoting transparency and long-term thinking—is linked to the last two actions in the Plan.

Action 9: Strengthen sustainability disclosures and accounting regulations. Corporate disclosures on sustainability issues are valuable for investors and stakeholders in making informed investment decisions, particularly concerning long-term value creation and a company's exposure to sustainability risks. Since 2018, an EU directive has required certain public interest entities—large listed companies with over 500 employees, banks, and unlisted insurance companies—to produce non-financial disclosures. These reports detail relevant information on key environmental, social, and governance issues and related risk management. However, this is often insufficient, and the disclosures tend to be too flexible and lacking in standardization. To address this, the Commission will act in four ways: (1) review the suitability of EU legislation related to corporate disclosures, including the Non-Financial Information (NFI) Directive; (2) review the guidelines on non-financial information by the second quarter of 2019; (3) establish a European corporate reporting laboratory for financial information in the third quarter of 2018 to promote best practices in sustainability reporting; and (4) encourage financial market actors to consider sustainability factors in their investment decisions and communicate them effectively.

There are also concerns in the accounting sector, as current accounting standards are not seen to encourage sustainable investment decisions. This concern arose after the adoption of the new accounting standard for financial instruments (IFRS 9) on October 6, 2016, which could have negative impacts on long-term investments. The Commission invited the European Financial Reporting Advisory Group to assess how the new or revised IFRS standards affect sustainable investments.

Action 10: Promote sustainable corporate governance and mitigate short-termism in capital markets. Corporate governance can contribute to achieving a more sustainable economy, enabling companies to develop new technologies or improve business models and performance. It can also facilitate job creation and stimulate innovation. However, a short-term focus on financial performance can lead to the undervaluation of sustainability-related risks and opportunities. To promote corporate governance that fosters sustainable investments, the Commission will conduct an analysis to decide whether: (a) to require company boards to develop sustainability strategies and measurable sustainability goals; and (b) to define rules obliging directors to act in the long-term interest of the company. Additionally, through the Action Plan and Action 10, the Commission aims to encourage a medium- to long-term outlook in capital markets. The Commission will ask European supervisory authorities, particularly ESMA, to assess whether capital markets unduly pressure companies to focus on short-term outcomes.

Having reviewed the ten actions of the European Commission's Action Plan, it is clear that this plan will make a significant contribution to achieving the goals of the Paris Climate Agreement and the United

Nations' Sustainable Development Goals, reinforcing Europe's ambition to lead the fight against climate change. However, the transition to a more sustainable economy and society undoubtedly requires coordinated global efforts. For this reason, the Action Plan can also serve as a model for future discussions in international forums, with the shared goal of promoting a more sustainable world. Indeed, sustainability approaches in non-EU countries show significant differences compared to European legislation.

The Action Plan

Two months after the adoption of the Action Plan on Financing Sustainable Growth, in May 2018, the European Commission developed a series of regulatory proposals, thereby initiating the implementation of the first measures outlined in the plan. These proposals position the EU financial system as a global leader in the transition towards a greener, cleaner, and more sustainable economy. Through these proposals, the full influence of the financial sector can be harnessed to combat climate change. There are compelling reasons to mobilize the financial sector in service of the planet's future. First, financial stability is already highly threatened by the effects of climate change (such as floods, land erosion, and droughts), which are causing severe economic damage. It is estimated that insured losses from such devastating events in 2017 amounted to approximately €110 billion, the highest ever recorded. Second, if we fail to address the severe environmental issues affecting the planet in time, many existing investments risk becoming obsolete. Lastly, the potential opportunities offered by sustainable economic activities must be fully leveraged. As stated in the European Commission's press release of May 24, 2018: "The EU financial sector has the potential to exponentially grow sustainable finance and become a world leader in this field. This should have a positive effect on economic growth and job creation, while supporting the Capital Markets Union's (CMU) objective of bridging finance with the needs of both the European economy and the EU's sustainability agenda."

Returning to the regulatory proposals, the key measures include:

> The establishment of a unified EU-wide classification system for environmentally sustainable activities ("taxonomy"). This proposal sets out harmonized criteria to identify environmentally sustainable

economic activities, allowing for the determination of the eco-sustainability of an investment, and outlines a process involving a multilateral platform to establish a unified EU classification system based on specific criteria. This system will determine which economic activities are considered sustainable. The Commission will identify activities that meet these criteria, taking into account current market practices and initiatives. This proposal will enable economic operators and investors to make more informed decisions. In short, this increased clarity will ensure that investment strategies target economic activities that genuinely contribute to sustainability objectives. Specifically, Article 3 [COM(2018) 353] of the legislative proposal states that an economic activity is considered environmentally sustainable, according to the taxonomy, if it substantially contributes to one or more environmental objectives without causing harm to the achievement of others.

Defining obligations for investors and disclosure requirements. This proposal aims to introduce greater consistency and clarity regarding how institutional investors incorporate ESG factors into the decision-making processes entrusted to them by their clients. It involves public disclosure obligations for institutional investors—such as asset managers, insurance companies, pension funds, and financial advisors—regarding how they integrate environmental, social, and governance (ESG) factors into their investments. Additionally, these actors must demonstrate how their investments align with ESG objectives and explain how they comply with the related obligations. This proposal addresses the need to resolve the current disorder in the European financial market, primarily caused by legislative differences across member states.

Creation of low-carbon investment benchmarks. This proposal responds to the need to clarify the methodology behind benchmark provision, making them more comparable and improving the decision making process for portfolio managers. Essentially, new categories of benchmarks will be developed, including a low-carbon benchmark and a positive carbon impact benchmark. This new market reference model should reflect the carbon impact of companies, providing investors with greater insight into the carbon footprint of a given investment portfolio.

Additionally, in May 2018, the Commission launched a public consultation on the integration of ESG criteria into the advice provided by investment firms and insurance product distributors to individual clients. This consultation seeks to reformulate the MiFID II and IDD Directives, as mentioned in Action 4 of the Action Plan. Essentially, when assessing whether a product meets a client's needs, these firms should also consider the client's sustainability preferences.

In June 2018, another important step was taken towards the implementation of the Action Plan: the European Commission appointed the Technical Expert Group (TEG) on Sustainable Finance, composed of 35 experts in sustainable finance from civil society, academia, the financial sector, and public bodies at both European and international levels. The role of the TEG is to provide support and advice to the Commission in implementing the Action Plan, focusing on several key topics closely related to the regulatory proposals just discussed. Specifically, the TEG's work covers:

> The European classification system, or "taxonomy," for environmentally sustainable economic activities (with a priority on environmental issues, especially climate change mitigation and adaptation).
>
> ESG disclosure of information on the policies and products of institutional investors.
>
> Improving guidelines for climate-related information reporting by large public-interest entities (listed companies, banks, asset managers, insurance companies).
>
> Common criteria for the development of climate benchmarks, establishing reliable reference standards to reduce the risk of greenwashing and increase market transparency.

In summary, the European Commission's strategy for implementing the Action Plan—aimed at redirecting capital flows towards sustainable investments and promoting greater transparency of information for market participants—rests on three key pillars: (a) the European taxonomy; (b) public disclosure obligations; and (c) climate benchmarks and sustainable investment standards. These three pillars will be further explored in the following sections through the study of the respective EU Regulations.

EU Regulation 2020/852: The Taxonomy Regulation

The definition of a European taxonomy for sustainable activities is arguably the most critical component of the European Union's strategy to achieve a more sustainable economy. Without a common definition of sustainability at the European level—for both financial and non-financial companies—it is difficult to imagine reaching the goals set by the EU. This taxonomy provides operators with a clear and reliable framework for making informed financial decisions. Additionally, the taxonomy helps mitigate the risks of greenwashing.

The key regulation that forms the basis of most considerations in this context is EU Regulation 2020/852, also known as the "Taxonomy Regulation," which came into effect on July 12, 2020. Before delving into the specifics of the regulation, which outlines the objectives and operating principles of the taxonomy, it's worth recalling its definition:

> The taxonomy is a classification of environmentally sustainable economic activities, designed as a tool to guide the investment choices of investors and companies towards the transition to economic growth that has no negative environmental impacts, particularly concerning the climate.

Thus, the European taxonomy equips financial markets and companies with a common standard for identifying sustainable activities. This need arises from the environmental—and particularly climate—objectives pursued by the European Union. In alignment with the Paris Agreement, the EU has committed to reducing greenhouse gas emissions by 55% by 2030 (compared to 1990 levels), aiming to achieve climate neutrality by 2050. Naturally, this economic and energy shift will require substantial resources, and beyond public funds, private capital will also play a crucial role. Estimates suggest that the EU will need €175–290 billion in additional annual investments. This brief analysis underscores the necessity of creating a taxonomy with clear, uniform rules to guide investors, combat greenwashing, and leverage private finance for sustainable economic activities.

Under EU Regulation 2020/852, the definition of environmentally sustainable economic activities—and the determination of an investment's level of sustainability—is based on the activity's ability to contribute to six environmental objectives, listed in Article 9 of the Regulation: "a) climate change mitigation; b) climate change adaptation; c) sustainable use and protection of water and marine resources; d) transition to a circular economy; e) pollution prevention and control; f) protection and restoration of biodiversity and ecosystems."

For an activity to be considered environmentally sustainable, it must meet four criteria (Article 3):

- Contribute substantially to one or more of the six environmental objectives.
- Do no significant harm to any of the other environmental objectives.
- Be conducted in compliance with minimum social safeguards (as outlined in OECD guidelines, UN documents on business and human rights, or ILO conventions).
- Comply with the technical criteria identified by the Technical Expert Group (TEG), which defines quantitative and/or qualitative requirements determining how an activity contributes to at least one of the six environmental objectives (Substantial contribution) and does not hinder the achievement of the other objectives (Doing No Significant Harm, DNSH).

Thus, to be included in the European Taxonomy, an economic activity must meet these four criteria.

Regarding the scope of the Regulation, Article 1 specifies that it applies to: "a) measures adopted by Member States or the Union that establish requirements for financial market participants or issuers in relation to financial products or corporate bonds marketed as environmentally sustainable; b) financial market participants offering financial products; c) companies subject to the obligation to publish non-financial or consolidated non-financial statements under Articles 19a or 29a of Directive 2013/34/EU of the European Parliament and the Council."

In this context, as emphasized by the Commission, the taxonomy serves as a practical, clear, and consistent guide for economic actors (policymakers, businesses, investors, etc.) on how to invest in activities that support a more sustainable economy. It is not, therefore, a mandatory list of environmental requirements.

As previously mentioned, the Taxonomy Regulation has been in force since July 12, 2020, following the publication of EU Regulation 2020/852 in the Official Journal of the European Union, which provides the general legal framework. However, the specific criteria and technical details, which establish the thresholds for defining an activity as sustainable, are set out in a series of Delegated Acts by the Commission, based on the reports produced by the TEG. It should be noted that the work of the TEG (Technical Expert Group on Sustainable Finance) concluded in September 2020, and it was replaced by the "Platform on Sustainable Finance."

For now, the Delegated Acts issued by the Commission specify and detail the technical criteria related to the first two sustainability objectives (climate change mitigation and climate change adaptation), while work on the remaining four objectives is still ongoing. In brief, the first Delegated Act relating to the climate aspects of the Taxonomy was published by the Commission in April 2021 and adopted on June 4 of the same year. This document outlines the technical criteria for identifying economic activities that can contribute substantially to the first two environmental objectives, namely climate change mitigation and adaptation. These criteria have been applicable since January 1, 2022, from which point financial operators must disclose whether their financial products align with the European Taxonomy. The April 2021 Delegated Act includes a list of over 100 economic activities from various sectors—including energy, transportation, construction, and manufacturing—and for each activity, it specifies technical criteria that indicate the quantitative thresholds for contributing to the mitigation and adaptation objectives. For the remaining four environmental objectives, a subsequent Delegated Act will be published, detailing the corresponding technical screening criteria.

It is important to note that the European Union's Taxonomy is considered the most advanced and comprehensive in the world, although other countries are also making progress in this area. However, in a globalized world, the risk is that if each country develops its own taxonomy system based on its national economy, the measures will be less effective and more challenging to harmonize. For example, the UK has a rapidly growing ESG market and is developing its own taxonomy, although it has stated that it will not follow the EU model. Similarly, Canada, Japan, China, and South Africa are working on unified systems to define green investments. However, particularly in the case of China, there are concerns about the stringency of its taxonomy and its willingness to align with European standards to foster sustainable finance development. Lastly, it appears that the United States is not yet inclined to adopt a taxonomy, and any criteria it adopts are unlikely to match the strict standards of the EU. A globally shared taxonomy would be the ideal solution, but given these circumstances, we risk waiting too long to collaborate at the international level.

The regulation UE 2019/2088: Sustainable Finance Disclosure Regulation *(SFDR)*

The European Regulation 2019/2088 (Sustainable Finance Disclosure Regulation, SFDR), introduced on November 27, 2019, and in force since March 10, 2021, sets new transparency obligations for financial market participants and financial advisors regarding the incorporation of sustainability risks into their investment processes. Essentially, the SFDR represents a crucial step in the EU's Action Plan for Sustainable Finance, with its primary goal being to regulate the ESG investment space and ensure greater clarity, transparency, and standardization of information for investors concerning ESG financial products. As a result, this regulation provides investors with a tool to compare financial products, evaluate their sustainability level, and make informed decisions aligned with their goals.

It is important to note that the SFDR, by standardizing reporting requirements for financial products, also aims to curb the growing issue of "greenwashing" in finance, which has reached alarming proportions.

The SFDR applies to all financial market participants (FMPs)—such as investment firms, banks, insurance companies, asset management companies, and pension funds—as well as financial advisors (FAs) based in the European Union or serving clients within the EU. The SFDR is a complex regulation consisting of 20 articles, designed to enhance the sustainability-related disclosures of financial service providers, both those who make investment decisions and those offering investment recommendations. These entities are required to disclose how they incorporate ESG factors in both their investment decision-making processes and the financial products they offer in EU markets.

The SFDR's key points revolve around two main aspects:

- Disclosure: the sustainability information that these entities must publicly disclose.
- Financial products: categorized into two ESG groups based on their characteristics.

Before diving into these two aspects, it's crucial to understand the definition of a "sustainable investment" or ESG. The SFDR provides a "qualitative" definition, listing—under Article 2, paragraph 17—three criteria that must be met for an investment to be considered sustainable: (a) an investment in an economic activity that contributes to an environmental or social objective, or in human capital or economically and socially disadvantaged communities; (b) such investment must not significantly harm any of the above objectives; (c) the enterprises benefiting from these investments must follow good governance practices.

Regarding disclosure obligations, these vary depending on the subject of the information (whether it refers to financial market participants/advisors or to the financial product) and the method of disclosure (whether via the web, pre-contractual, or periodic reporting). For financial products, the regulation mandates that pre-contractual and periodic documents must include:

How sustainability risks are defined, measured, and integrated into investment decisions, along with an indication of their potential impact on the financial performance of the product.

For investments labeled as "sustainable," the specific sustainability objectives pursued and the methods for achieving them must be disclosed.

For investments not categorized as ESG, an explanation must be provided as to why sustainable investments were not considered as part of the investment options.

It is important to clarify that the SFDR does not require entities to direct capital flows toward sustainable investments. Instead, it mandates that they inform clients about this category of investments, helping clients make more conscious choices. In other words, entities that do not offer sustainable investments must provide a clear explanation: while not directly imposing ESG investments, the SFDR indirectly promotes them.

Financial market participants and advisors must disclose to end investors, through their websites or promotional materials, the following information:

Details on their policies for integrating sustainability risks into investment decision-making or advisory services (similar to the disclosure for financial products mentioned above).

The potential negative impacts that investment decisions or advice might have on sustainability factors.

How their remuneration policies align with the integration of sustainability risks.

In addition to disclosure, the SFDR's other key element is the classification of financial products, which, as previously mentioned, are divided into two distinct ESG categories:

Article 8 defines the first category of ESG investment products as those that "promote, among other characteristics, environmental or social characteristics, or a combination of those characteristics."

Article 9 outlines the second category, which includes any financial product that "has sustainable investments as its objective" (with or without a reference index) or any financial product that "aims to reduce carbon emissions."

Financial products that do not fall into these categories, are not focused on sustainability, or do not meet the conditions set out in Article 2, paragraph 17, cannot be labeled as "sustainable investments" or ESG.

More than a year after the SFDR came into effect, it is clear that this regulation is essential for fostering a European market for sustainable products, which have seen increasing interest in recent years. This is confirmed by a recent analysis from Morningstar (as of December 31, 2021), which indicates that assets in funds classified under Articles 8 and 9 of the SFDR account for 42.4% of all European funds, reaching €4 trillion by the end of 2021.

However, since transparency in disclosure alone cannot radically transform markets, the European Union continues to work towards a more sustainable economy. It should be noted that non-EU countries have adopted different regulations regarding sustainability disclosure, with each country developing its own standards, some more aligned with the European framework than others. However, as these regulations are relatively new, their concrete effects will be visible later compared to the standards set by the European Union.

Climate Indices and Sustainable Investment Standards
Climate indices and sustainable investment standards are essential tools in guiding the transition towards a low-carbon economy. These indices provide benchmarks for measuring the environmental impact and sustainability of investments, helping investors assess the alignment of their portfolios with climate goals. Sustainable investment standards, on the other hand, establish guidelines and criteria that define what qualifies as a "sustainable" investment. They ensure transparency, accountability, and consistency in how companies and financial products are evaluated in terms of their environmental, social, and governance (ESG) performance.

Climate indices, such as the MSCI Climate Index or the FTSE Russell Green Revenues Index, evaluate companies based on their environmental impact, particularly their carbon footprint, energy consumption, and greenhouse gas emissions. These indices offer investors a way to track the performance of companies that are actively working to reduce their environmental impact, and they provide a reference point for investors looking to prioritize climate-friendly investments.

Sustainable investment standards, such as the Global Reporting Initiative (GRI) and the Principles for Responsible Investment (PRI), create frameworks for evaluating the ESG aspects of an investment. They require companies to disclose detailed information on their sustainability practices, ranging from environmental impact to social responsibility and governance structures. By adhering to these standards, companies can demonstrate their commitment to sustainability, while investors can make more informed decisions about where to allocate their capital.

These tools play a pivotal role in driving capital towards more sustainable projects and ensuring that investments contribute to long-term environmental goals, such as reducing global warming and achieving carbon neutrality.

THE CIRCULAR ECONOMY AND THE NEW EUROPEAN INDUSTRIAL STRATEGY

Having examined the core principles of the European Commission's 2018 Action Plan for Financing Sustainable Growth, it is essential to ask: how can these sustainability best practices be effectively integrated into the social and entrepreneurial fabric of the European economy? After exploring these aspects, we will focus on the relationship between businesses and banks—an area central to this monograph—in the final sections of this chapter.

To understand this integration, it is crucial to explore the "circular economy" concept, which contrasts sharply with the current linear economic model. The circular economy is defined as "a model of production and consumption that involves sharing, lending, reusing, repairing, refurbishing, and recycling existing materials and products for as long as possible" (Geissdoerfer et al., 2017). On March 11, 2020, the European Commission launched the Circular Economy Action Plan, targeting sectors for immediate intervention to accelerate the shift to a sustainable economy. This comprehensive program is integral to the European

Green Deal of December 12, 2019, which seeks to make Europe the first climate-neutral continent by 2050 by promoting resource efficiency, reducing pollution, and fostering a circular economic model (European Commission, 2020).

The Circular Economy Action Plan focuses on consumer products, emphasizing durable, repairable designs with higher recycled content, and promotes models like "product-as-a-service" to curb single-use products (Korhonen et al., 2018). Priority sectors include electronics, ICT, batteries, packaging, textiles, construction, food, and nutrients—areas with high resource intensity and circularity potential. For example, the plan promotes repairable, eco-friendly devices, sustainable packaging, and recycling-focused measures in sectors like food and construction (Ness, 2008). The Circular Economy Action Plan also addresses waste reduction, targeting recycling improvements through the extended producer responsibility scheme and updated waste sorting policies (Prieto-Sandoval et al., 2018).

Sustainability must be embedded in key sectors. For food, this entails waste reduction and sustainable dietary choices, while in transport, it involves promoting cleaner mobility systems. In construction, the focus is on improving building energy efficiency and incorporating circular principles in design (de Jesus & Mendonça, 2018). In February 2021, the European Parliament reinforced this strategy, calling for a fully circular economy by 2050 and new legislative measures for construction, textiles, and packaging introduced in 2022.

The European Industrial Strategy, presented in March 2020, aligns with these goals by supporting industry in leading climate neutrality and digital advancement. This strategy is built around seven pillars: enhancing the digital single market, ensuring global competitiveness, supporting climate neutrality, fostering circularity, driving innovation, advancing skills, and investing in the green transition (D'Amato et al., 2019). This transformative approach emphasizes collaborative efforts across EU institutions, member states, and industry stakeholders to create a robust market for clean technologies and position European industry as a global leader in sustainability.

The regulatory framework for sustainability in Europe is evolving rapidly, spurred by the increasing urgency of environmental concerns and the visible consequences of inaction. Within the European context, proactive and integrated approaches to sustainability are becoming a fundamental necessity.

Non-financial Declaration, Disclosure Obligations, and Sustainability Reporting: From the NFRD Directive to the CSRD

The increasing focus on sustainability, encapsulated in ESG (Environmental, Social, and Governance) principles, has deeply influenced corporate strategy, with companies now factoring ESG considerations into strategic decisions (Eccles et al., 2014; Serafeim, 2020). This shift reflects a broader change from purely economic and financial priorities to an integrated view of value creation that includes environmental and social dimensions (Khan et al., 2016). It is no longer sufficient to consider only costs and revenues in evaluating organizational performance; factors such as environmental impact, workforce well-being, stakeholder relations, natural resource management, and community engagement are increasingly central to understanding a company's long-term viability and reputation. As companies adopt ESG-focused strategies, the importance of non-financial performance reporting tools has surged, complementing traditional financial statements with insights into sustainability-related efforts (Friede et al., 2015).

Non-financial reporting has emerged as a key indicator of ESG integration within companies, and it exists in various forms. Voluntary reports, such as Environmental Reports, Social Reports, and Sustainability Reports, each provide different lenses through which stakeholders can assess a company's impact. Environmental Reports, for instance, detail a company's strategies and results concerning environmental protection, aiming to inform stakeholders of its ecological footprint (Moser & Martin, 2012). Social Reports outline the company's social impact on its surrounding community, while Sustainability Reports encompass environmental, social, and governance aspects in one document, offering stakeholders a holistic view of the company's responsibility practices (Ioannou & Serafeim, 2019).

The Integrated Report presents a consolidated approach, linking financial and non-financial performance to illustrate how a company's strategy, governance, and external impacts create value over time (Adams, 2015). Unlike Sustainability Reports, Integrated Reports structure information around distinct "capitals"—such as financial, productive, human, social, and natural—providing a comprehensive framework for evaluating value creation across multiple dimensions (Eccles & Krzus, 2010). Meanwhile, the Non-Financial Statement (NFS), or Non-Financial Disclosure (NFD), is a mandatory report under the European Union's Directive 2014/95/EU (NFRD) for large companies and public-interest entities. The NFRD requires disclosure on environmental, social, employee-related, human rights, anti-corruption, and bribery issues, helping stakeholders assess the broader impacts of corporate operations (European Commission, 2014).

Further regulatory developments have deepened the requirements for non-financial reporting. The EU Taxonomy Regulation (2020) mandates that companies disclose their alignment with eco-sustainable activities, focusing on revenue, capital expenditures, and operating expenditures tied to sustainable practices (Pizzi et al., 2021). The European Union also implemented Directive 2013/34/EU and its amendment, Directive 2014/95/EU, requiring companies to provide a Non-Financial Statement, addressing issues such as environmental impacts, human rights, and anti-corruption efforts (European Commission, 2014). In the Italian context, Legislative Decree No. 254 of 2016 operationalizes these EU directives, mandating that listed companies and other public-interest entities disclose non-financial information on ESG aspects (Di Tullio & Pascale, 2020).

The regulatory landscape continues to evolve. In April 2021, the European Commission proposed the Corporate Sustainability Reporting Directive (CSRD) to address gaps in the NFRD by expanding disclosure requirements to all large companies and listed SMEs, reinforcing the "double materiality" principle, and mandating third-party assurance of sustainability information (European Commission, 2021). Approved in 2022, the CSRD aims to reduce greenwashing, enhance accountability, and align corporate disclosures with the EU's sustainability goals (European Financial Reporting Advisory Group, 2021).

In conclusion, the regulatory framework surrounding corporate sustainability reporting has become an essential element in the transition to a sustainable economy. By providing accurate, reliable, and comparable ESG information, these frameworks support the goals of the European Green Deal and the EU's Sustainable Finance Action Plans, ensuring that financial markets can make informed decisions to support the green transition (European Commission, 2019; HLEG, 2018).

The Banking Sector: EU Regulatory and Supervisory Paradigm on ESG Factors

The increasing importance attributed to ESG factors has significantly influenced the relationship between businesses and banks. As discussed in Chapter 1 (Sect. 1.8), banks, due to their intermediary role, play a critical part in driving the transition toward a more sustainable economy and growth model.

Banks can either promote or hinder sustainable (or unsustainable) behaviors of states, companies, and individuals. Furthermore, these financial institutions can trigger structural changes in society. For example, banks may consider sustainability aspects in investment decisions, such as in the case of Sustainable and Responsible Investments (SRI). They may also apply negative criteria (e.g., excluding nuclear sectors as clients) or positive criteria (e.g., favoring the development of renewable energy sectors) in their decision-making process, directly influencing corporate priorities and values. Sustainability criteria can also be applied in lending decisions. Given their substantial impact on society and the environment, banks are increasingly addressing environmental and social challenges by adopting new sustainable practices.

By incorporating ESG criteria, banks can lead the economic transition, creating a new credit culture and guiding companies toward achieving a more robust and sustainable economic system. It is crucial to avoid repeating the mistakes of the past, particularly those that led to the 2007 global financial crisis, which was followed by a severe credit crisis for non-financial companies and a devastating economic recession. The persistence of the recession increased credit risk, while fragile debt recovery procedures, long recovery times, and partial collections fueled the rise of non-performing loans (NPLs), which, in turn, worsened bank balance sheets. Therefore, it is essential to closely examine the risks related to credit provision in the coming years and reduce the stock of NPLs to improve the quality of banking assets. This shift represents a move from managing existing bad debt (a reactive approach) to proactively managing credit even before it is recorded in the balance sheet, a direction advocated by European regulators. This approach aims to strengthen financial sustainability, fortifying the banking system and its resilience.

In recent years, the banking sector has seen increasing attention to ESG factors, involving regulators, supervisory authorities, and institutions.

Similar to companies, banks are also being guided by European supervisory authorities through guidelines and regulations aimed at steering the financial system towards the sustainability objectives set by the EU. A significant development in this area came in 2022 when the European Banking Authority (EBA) published its Action Plan on Sustainable Finance (EBA, 2022). This document outlines how the EBA intends to integrate sustainability principles into the European banking and financial sector and incorporate ESG factors and risks into financial services.

The EBA's primary objectives regarding sustainable finance are to support the short-, medium-, and long-term stability of the financial system, improve prudential regulatory frameworks to facilitate sustainable operations, and equip institutions with tools to monitor and evaluate ESG risks in their supervisory practices. The EBA's Action Plan provides an overview of guidelines and standards to help banks and supervisory authorities assess ESG factors and risks.

The EBA's mandates and responsibilities concerning ESG factors and risks are not only defined by the European Commission's Action Plan but are also enshrined in four legislative acts. These include:

> Considering sustainable business models, developing monitoring systems to evaluate potential inclusion of ESG risks in supervisory processes, and creating common methodologies for these purposes. Complying with Article 449a of the EU Regulation 2019/876 (Capital Requirements Regulation—CRR), which mandates large listed financial institutions to disclose ESG risks, including physical and transition risks, from June 28, 2022.
>
> Assessing whether a specific prudential treatment for exposures related to sustainable activities or assets is justified.

Following the European Commission's Action Plan to promote sustainable corporate governance and mitigate short-term pressures from capital markets on companies, with the EBA tasked with investigating and addressing these concerns for firms considering long-term risks and opportunities, including environmental and climate risks.

To fulfill these mandates, the EBA has adopted a four-phase sequential approach:

Strategy and risk management: This phase involves developing proposals for managing ESG risks and integrating them into governance, risk management, and supervisory procedures. The EBA is working towards establishing a clear definition of ESG risks and creating criteria and methods to assess the impact of these risks on credit institutions.

Key metrics and disclosure: The EBA is developing technical standards to implement disclosure requirements for banks, focusing on both qualitative and quantitative metrics, such as the Green Asset Ratio, to measure and monitor ESG risks.

Stress testing and scenario analysis: This phase entails developing climate change stress tests to assess banks' vulnerability to climate and environmental risks, with the goal of creating methodologies to measure the impact of economic scenarios on financial institutions.

Prudential treatment: This phase requires the EBA to evaluate whether prudential treatment of exposures related to assets with significant environmental and social objectives is justified. A report on this is due by June 2025.

These four pillars of the EBA's 2019 Action Plan on Sustainable Finance represent key steps in the integration of ESG factors in the banking sector, encouraging institutions to adopt concrete measures.

In addition to the EBA's initiatives, the European Central Bank (ECB) has also contributed to sustainable finance, notably by revising prudential expectations concerning provisioning for new non-performing loans. European banks are now expected to provide detailed information on the quality and risk of new loans, especially focusing on the relationship between pricing and credit risk. The ECB has also published its Guidelines on Loan Origination and Monitoring, aimed at improving governance, risk management, and loan origination processes, while incorporating ESG factors into lending practices.

Moreover, the ECB published its Guide on Climate and Environmental Risks in November 2020, outlining expectations for banks regarding the prudent management of climate and environmental risks. Although not binding, this guide serves as a foundation for supervisory dialogue and is part of the broader effort to prepare the banking sector for managing climate and environmental risks as genuine financial risks.

Finally, the EBA's June 2021 report on managing and supervising ESG risks provides comprehensive recommendations on how to integrate ESG factors and risks into credit institutions and investment firms. This report emphasizes the importance of considering the potential financial impacts of ESG risks over time and highlights the need for proactive and forward-looking assessments.

In conclusion, the growing focus on ESG factors in the banking sector reflects a broader shift towards financial sustainability, with the potential to yield significant benefits for banks. By supporting more sustainable companies, banks are likely to invest in stronger, more resilient businesses, ensuring safer long-term returns. Furthermore, by fostering a new credit culture, banks will be better equipped to understand and manage risks, enhancing financial stability and reducing the accumulation of non-performing loans. In short, by integrating ESG factors, banks will gain both economic and intangible advantages, just as other companies complying with sustainability factors do (Fig. 2.2).

Additionally, to address ESG risks effectively, the EBA, in its 2021 report on ESG risk management and supervision, provided key guidelines on qualitative and quantitative indicators, as well as methodological

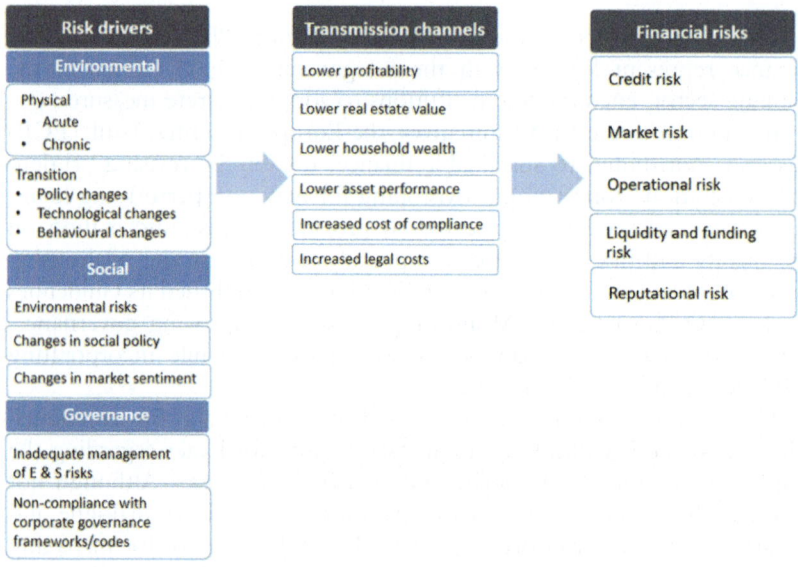

Fig. 2.2 Overview of ESG risk factors, their transmission channels, and their impact on financial risk categories (*Source* EBA, *EBA/REP/2021/18*, p. 34)

tools to assess the financial impact of ESG risks. Establishing common ESG indicators and methodologies (in addition to defining ESG risks and factors) is essential to promote the integration of sustainability considerations into financial decision-making and supervisory processes. This also ensures greater transparency and better information for relevant stakeholders, while helping to prevent greenwashing risks. The EBA notes that while many institutions and supervisory authorities have begun incorporating ESG factors into their operations, the practice of ESG risk assessment is still in its early stages.

There are several challenges or critical issues in integrating ESG risks into management processes and supervisory practices:

- Uncertainty regarding the timing and effects of policies and regulatory interventions (which fall under the competence of each EU Member State). These details are often difficult to predict, as are the timing and effects of physical risks.
- A lack of relevant, comparable, reliable, and user-friendly ESG data (especially for small and medium-sized enterprises, or SMEs).
- Methodological limitations, as many risk management models rely on historical data to estimate current or future risks, and ESG factors are often not adequately reflected in such data.
- A misalignment between the short-term focus of traditional management tools and the long-term materialization of ESG risks.
- The multi-faceted impact of ESG risks on institutions, as these risks can affect various categories of financial risk and thus impact institutions' financial positions in multiple ways. These impacts should be assessed for each category of financial risk and across multiple categories.
- The non-linear nature of ESG risks, which can lead to complex chain reactions and cascading effects, potentially triggering unpredictable socio-economic dynamics.

To address these challenges, the EBA outlines three fundamental phases for institutions to implement. The first phase is 'Identification,' which involves classifying activities and exposures based on their ESG characteristics to facilitate the identification of the main potential drivers of ESG risks using specific qualitative and quantitative indicators. In this phase, institutions' exposures can be categorized by asset class, sector, counterparties, geography, duration, or stage in the asset lifecycle, to identify activities or exposures most vulnerable to ESG risks.

Once exposures are classified, the second phase is 'Evaluation,' which involves using methodological tools to measure, assess, or estimate the potential impact of ESG risks on institutions' exposures. This phase helps institutions gain a better understanding of their financial vulnerability to ESG risks.

The final phase is 'Action,' which calls for the incorporation of ESG risks into risk management by adopting corporate strategies and risk management approaches that support the monitoring and control of these risks.

After submitting this report to the European Parliament, the Council, and the Commission, the EBA encourages institutions to actively consider its recommendations. To be adequately prepared for ESG-related challenges and regulatory requirements, timely and proactive actions are essential.

Concluding the discussion on the banking sector's role in transitioning towards a more sustainable economy, it is an important to highlight a recent initiative by the European Banking Authority (EBA), which, on December 13, 2022, published a new Sustainable Finance Roadmap. This roadmap outlines the goals and action plan for implementing its mandates and tasks related to sustainable finance and ESG risks. The roadmap presents the EBA's sequential approach over the next three years, aiming to incorporate ESG risk issues into the banking sector while supporting the EU's broader efforts towards a more sustainable economy. Notably, this roadmap updates and replaces the EBA's previous 2019 Action Plan on Sustainable Finance, ensuring continuity while adapting to market developments and regulatory changes. The EBA's new mandates extend beyond the core pillars of the banking system—market discipline, supervision, and prudential requirements—to cover other areas of sustainable finance and ESG risks. However, these mandates align with the EBA's broader goal of fostering the stability, resilience, and proper functioning of the financial system.

The roadmap focuses on eight main areas of intervention, which are summarized in the figure below (Fig. 2.3).

First, in the area of "transparency and disclosure" (1), the EBA aims to continue its efforts in developing and implementing ESG risk frameworks for institutions, while also expanding sustainability-related disclosure. Regarding "risk management and supervision" (2), the agency plans to

Fig. 2.3 Key points of the EBA's sustainable finance roadmap (*Source* EBA, *EBA/REP/2022/30*, p. 9)

further ensure that ESG factors and risks are properly integrated into the risk management and supervisory frameworks of institutions, including through new initiatives such as climate stress tests. In the area of "prudential regulation" (3), the EBA has initiated an assessment phase to analyze and, if necessary, adjust the current prudential treatment of banks' exposures to incorporate environmental and social considerations.

Another key aspect of this EBA roadmap on sustainable finance is the development of stress tests related to banks' exposure to ESG risks, aimed at identifying the most significant vulnerabilities within banks ("stress tests," 4). Additionally, the EBA will work on developing standards and labels to qualify green banking products ("standards and labels," 5). The roadmap also aims to mitigate greenwashing risks (6) by adopting measures to better identify and manage such risks through more effective controls. Furthermore, the EBA is committed to integrating ESG risk information within a new supervisory reporting framework ("supervisory reporting," 7).

Lastly, the final point of the roadmap ("ESG risks and sustainable finance monitoring framework," 8) highlights the EBA's intention to actively evaluate and monitor developments in sustainable finance and the ESG risk profiles of institutions in the coming years.

The issues discussed in this section illustrate the path toward a more sustainable economy, one that the banking sector is also preparing for. Banks, like other businesses, are key players in leading this transition by increasingly incorporating sustainability topics into their policies and strategies. In response to market developments and the regulatory requirements discussed, ESG factors now hold significant influence within financial institutions, which have embraced this challenge and are investing considerable efforts in this area. As observed, these topics are constantly evolving, and we can expect stricter ESG regulations and guidelines in the coming years, which will shape the future state of society, the economy, and the health of the planet.

CHAPTER 3

ESG and the Cost of Debt: An Empirical Analysis of Sustainability in Bank Financing

Abstract This chapter investigates the relationship between corporate ESG (Environmental, Social, and Governance) performance and the cost of debt (CoD) incurred by firms. Building on prior research, it explores whether strong ESG practices lower perceived credit risk, resulting in favorable borrowing terms from banks. The study employs ESG scores from the Bloomberg Terminal and financial data from Orbis, analyzing a sample of 107 EU-based companies across capital-intensive and human capital-focused industries over five years (2017–2021). The findings suggest a significant negative relationship between ESG performance and CoD, with higher ESG scores correlating with reduced borrowing costs. Control variables such as profitability, firm size, and financial leverage were also assessed. Results highlight the growing integration of ESG metrics into credit risk evaluations, underscoring the role of sustainable finance in incentivizing corporate responsibility.

Keywords ESG performance · Cost of debt (CoD) · Sustainable finance · Corporate credit risk · Bank loans · ESG integration

© The Author(s), under exclusive license to Springer Nature
Switzerland AG 2025
N. Del Sarto, *ESG Factors and Financial Outcomes in Banks*,
https://doi.org/10.1007/978-3-031-87748-3_3

95

INTRODUCTION

In recent years, the importance of sustainability issues, encapsulated in the ESG (Environmental, Social, and Governance) framework, has grown so significantly that it is now integral to corporate, governmental, and financial agendas. This shift is driven by urgent environmental concerns, such as climate change and resource depletion, as well as social and governance pressures for transparency and accountability (Khan et al., 2016; Schoenmaker & Schramade, 2019). The transition to a sustainable, green growth model has become essential for both economic and societal stability, as organizations increasingly recognize the interconnectedness of financial success and environmental and social responsibility (Clark et al., 2015). ESG factors are now central to the operations of businesses, governments, financial institutions, and all market participants, reflecting a more holistic approach to value creation (Eccles & Klimenko, 2019).

As ESG considerations become increasingly embedded in business practices, their influence on financial decisions has grown. Notably, ESG practices are gaining traction in the realm of debt financing, where investors and banks alike are beginning to assess companies' sustainability profiles in determining creditworthiness and lending terms (Bauer & Hann, 2010; Goss & Roberts, 2011). This study investigates whether companies with strong ESG performance receive preferential treatment in terms of borrowing costs from banks. Specifically, we explore whether high ESG compliance leads to a lower cost of debt (CoD), hypothesizing that banks may perceive ESG-compliant firms as lower-risk borrowers. Such preferential treatment could manifest as reduced interest rates and better loan conditions, based on the rationale that firms prioritizing sustainable practices present less financial risk due to improved reputational standing, operational efficiency, and resilience to regulatory and environmental disruptions (Hoepner et al., 2016; Nandy & Lodh, 2012).

Our central research hypothesis posits a negative relationship between ESG factors and CoD, suggesting that higher ESG scores are associated with lower borrowing costs. This hypothesis is grounded in the belief that ESG compliance reduces risks perceived by lenders, potentially resulting in more favorable loan terms (Chava, 2014). Supporting this hypothesis, studies indicate that companies with robust ESG practices tend to exhibit improved operational stability and stakeholder trust, which can translate into lower credit risk and more favorable borrowing costs (Devalle et al., 2017). However, while empirical studies have confirmed the positive impact of ESG on equity financing, evidence regarding the influence of ESG on debt financing is mixed, with some studies finding no significant relationship or even suggesting that ESG practices are not yet fully integrated into credit evaluations (Cooper & Uzur, 2015; Goss & Roberts, 2011).

The importance of understanding the relationship between ESG factors and CoD is underscored by the growing emphasis on sustainable finance, which aims to align capital allocation with environmental and social goals (Palmieri, & Geretto, 2024). As the financial sector seeks to address broader societal challenges, banks and investors are increasingly expected to reward companies that actively pursue sustainable business models, not only through equity markets but also in debt markets (Friede et al., 2015; Schoenmaker, 2017). This study contributes to the literature by addressing the gap in empirical research on ESG's impact on debt financing, examining whether sustainable practices yield tangible financial benefits in terms of reduced borrowing costs.

Theoretical Background: Bank Loans and ESG

The relationship between corporate ESG (Environmental, Social, and Governance) performance and the cost of bank loans remains relatively underexplored in academic literature, with most studies focusing on the influence of ESG factors on equity financing or bond issuance rather than bank loans (Chava, 2014; Liang & Renneboog, 2017). Historically, research has primarily examined how ESG affects the cost of equity and corporate financial performance, with relatively less emphasis on the impact of ESG on the cost of debt, including bank loans (Attig et al., 2013).

For example, Devalle et al. (2017) highlight that a general consensus exists in the literature regarding ESG's positive impact on reducing the cost of equity, as higher ESG scores are often linked to lower equity costs due to improved risk profiles. This is supported by Friede et al. (2015), who provide meta-analytical evidence that ESG factors are generally associated with superior financial performance, reinforcing the argument that sustainability reduces risk exposure. However, this agreement does not fully extend to the cost of debt, where research is still limited and findings are inconsistent.

Some studies find a neutral or even negative association between ESG and bank loan costs. Goss and Roberts (2011), for instance, suggest that lenders may not always reward companies for sustainable practices, sometimes charging higher interest rates, possibly due to perceived risks or costs associated with compliance. Similarly, Oikonomou et al. (2014) point out that ESG-related controversies can increase firms' credit spreads, particularly when governance issues are perceived as a significant risk factor. Conversely, Cooper and Uzur (2015) and Hoepner et al. (2016) find evidence that strong CSR (Corporate Social Responsibility) practices correlate with reduced borrowing costs, indicating that firms committed to sustainability are viewed as lower risk. Nandy and Lodh (2012) further support this view, noting that companies with robust ESG practices often benefit from lower loan negotiation costs, suggesting that ESG compliance can positively influence loan terms. Additionally, Albuquerque et al. (2019) argue that ESG leaders often enjoy enhanced reputational benefits, which translate into reduced capital costs, including lower debt premiums (Gangi et al., 2021).

The literature generally agrees that companies with strong ESG performance tend to perform better financially, benefiting from enhanced stakeholder loyalty and reduced vulnerability to negative events, which ultimately can improve loan conditions (Deng et al., 2013). This aligns with modern theories of stakeholder capitalism, as articulated by Freeman et al. (2004), which emphasize that firms managing ESG issues effectively create long-term value for both shareholders and creditors. Furthermore, banks increasingly factor ESG considerations into their credit assessments. While some may avoid lending to firms with poor ESG profiles due to reputational concerns, others may adjust interest rates to offset perceived risks (Bauer & Hann, 2010). This reflects the growing trend of integrating non-financial metrics into credit risk evaluation, a practice supported by the Principles for Responsible Banking established by the United Nations Environment Programme Finance Initiative (UNEP FI, 2019).

Additionally, regulatory shifts and market expectations are accelerating this trend. The European Union's Taxonomy for Sustainable Finance and the Sustainable Finance Disclosure Regulation (SFDR) mandate that financial institutions assess ESG factors more rigorously in their lending and investment practices (European Commission, 2021). These frameworks aim to standardize ESG evaluation, potentially reducing inconsistencies in how ESG impacts credit terms. Furthermore, the alignment of lending practices with global initiatives, such as the Paris Agreement, reinforces the role of banks as critical drivers of sustainable development (Berg et al., 2020).

Despite these advancements, the relationship between ESG performance and the cost of debt remains complex, as it is influenced by sectoral and geographical differences. For instance, larger firms or those in heavily regulated industries may experience greater scrutiny of their ESG practices, resulting in more pronounced effects on loan terms (Cheng et al., 2014). Meanwhile, smaller firms, often lacking formal ESG reporting frameworks, may face challenges in translating sustainability efforts into financial benefits.

In light of these mixed findings, this study aims to fill the gap in existing research by analyzing how corporate ESG performance impacts the cost of bank financing. Specifically, we investigate whether firms adhering to ESG principles can achieve more favorable borrowing terms through reduced debt costs, contributing to the broader understanding of ESG's influence on corporate financing structures. Furthermore, this analysis considers how ESG's three pillars—environmental, social, and governance—individually influence bank lending, addressing the heterogeneity in existing findings and paving the way for more targeted applications of ESG criteria in financial decision-making.

METHOD

The empirical analysis presented in this final chapter seeks to determine whether the sustainability (ESG) policies adopted by firms impact the cost of debt as determined by banks. Previous studies suggest that ESG practices may influence credit conditions, as banks increasingly consider a firm's ESG profile in assessing risk and pricing loans (Bauer & Hann, 2010; Goss & Roberts, 2011). Our hypothesis is that companies with robust ESG policies are perceived as lower-risk by financial markets, leading to favorable loan terms, such as a reduced cost of debt. This expectation aligns with the broader understanding that ESG practices enhance a firm's reputation and operational stability, thus mitigating risk and potentially lowering borrowing costs (Hoepner et al., 2016; Nandy & Lodh, 2012).

To account for the delayed impact of ESG policies, we incorporate a one-year lag in the independent variable "ESG factors," based on the assumption that sustainability initiatives require time to influence financial outcomes and are unlikely to yield immediate effects within the same fiscal year. The selection of a one-year lag is grounded in both theoretical and practical considerations. Financial institutions and other stakeholders typically need time to assimilate ESG-related information and integrate it into their credit evaluations, as this process involves analyzing disclosures, assessing risks, and adjusting lending terms accordingly (Chava, 2014; Devalle et al., 2017).

The decision to use a one-year lag was also informed by preliminary testing, where we examined different lag periods, including two- and three-year lags, to evaluate their explanatory power. While longer lags did show some effects, the one-year lag provided the strongest and most consistent results, aligning with prior studies that suggest ESG benefits often become apparent within a short- to medium-term time frame (Liang & Renneboog, 2017). This approach strikes a balance between capturing the delayed impact of ESG initiatives and avoiding overly extended periods that might dilute the relevance of the findings due to confounding factors.

By explicitly incorporating this lagged structure, we address both the temporal dynamics of ESG impact and mitigate concerns about reverse causality, ensuring that the observed relationship reflects the influence of ESG performance on debt costs rather than vice versa. This methodological choice enhances the robustness and interpretability of the results, contributing to a more precise understanding of the interplay between sustainability initiatives and financial outcomes.

The subsequent sections provide a detailed description of the sample, key variables, and statistical model employed to test the hypothesis. By doing so, we aim to offer a comprehensive analysis of the relationship between corporate ESG performance and bank loan pricing, addressing a relatively underexplored area within the financial literature.

DATA

The research hypothesis was tested on a sample of companies selected from the "Orbis" database, the flagship corporate database of Bureau van Dijk (a Moody's Analytics company). Orbis is considered one of the largest databases, offering detailed and comparable information on over 450 million companies worldwide. This database was used to create the sample and to retrieve the financial data necessary for the analysis.

The empirical analysis covers a five-year period, focusing on the financial statements available from 2017 to 2021. While these are not the latest available years, this specific period was chosen because it includes the years leading up to and during the COVID-19 pandemic. This period is particularly interesting due to the financial difficulties faced by firms and the critical role ESG performance played in enhancing firm resilience during the crisis. As highlighted by Gao and Geng (2024), ESG factors were instrumental in mitigating risks and supporting companies in navigating unprecedented challenges during the pandemic. The data for these five fiscal years was extracted to capture these dynamics and provide a robust context for the analysis.

The following search criteria were used within Orbis to construct the company dataset:

Company status: Active companies only.

Geographic area: European Union [27]. This criterion includes Austria, Belgium, Bulgaria, Croatia, Cyprus, Czech Republic, Denmark, Estonia, Finland, France, Germany, Greece, Hungary, Ireland, Italy, Latvia, Lithuania, Luxembourg, Malta, Netherlands, Poland, Portugal, Romania, Slovakia, Slovenia, Spain, and Sweden. The choice to focus on the EU is linked to the monograph reference to the European Union's regulatory framework.

Company type: "Companies" were selected, excluding banks, financial institutions, and insurance companies, as is common in most literature on these topics (Devalle et al., 2017). Publicly listed companies were selected, given the broader availability of data for these entities.

Financial profile and number of employees: To segment the sample, companies with a minimum of 200 employees and a minimum total balance sheet of USD 20 million in the years under analysis (2017–2021) were selected. This was done because these entities are subject to the NFRD Directive, and data availability is higher for such firms.

Financial year-end: December was selected to ensure that financial statements and data were comparable across the sample.

Industry sectors: The focus was on the manufacturing sector, selecting two capital-intensive sectors, "Metallurgy and Metal Products" and "Chemicals, Pharmaceuticals, Petroleum, Rubber, and Plastic Products," as these industries are involved in heavy product manufacturing. Two additional sectors with a strong human capital component, "Textile and Apparel" and "Food and Tobacco," were also included. These sectors were chosen because, as discussed in the previous chapter (Sect. 2.5), they represent areas where the EU intends to intervene swiftly to accelerate the transition to a more sustainable economy. These sectors are among the largest users of raw materials, greenhouse gas emitters, and waste producers.

After applying these criteria, the final sample comprised 288 companies, resulting in 1,440 observations. The sector breakdown is as follows: 61 companies in "Metallurgy and Metal Products," 126 in "Chemicals, Pharmaceuticals, Petroleum, Rubber, and Plastic Products," 33 in "Textile and Apparel," and 68 in "Food and Tobacco."

Difficulties arose in collecting ESG data for the sample companies due to limitations in data availability. The Orbis database does not provide historical ESG information, as it only began predicting ESG scores in 2021. Consequently, the Bloomberg Terminal was selected as the primary source for ESG data retrieval. Bloomberg was preferred over other ESG data providers because of its comprehensive coverage of publicly listed companies, its established reputation for reliable and standardized financial and ESG metrics, and its alignment with the research objectives. Bloomberg's ESG scoring methodology emphasizes transparency, consistency, and comparability, making it particularly suitable for examining the relationship between ESG performance and financial outcomes.

Despite its strengths, ESG scores were not available for all companies in the sample, even among publicly listed firms. Some companies either did not report sufficient ESG-related data or had gaps in their disclosures, leading to incomplete scores. Bloomberg's reliance on reported data ensures methodological rigor but is limited by the disclosure practices of individual firms. These limitations were carefully considered in the analysis, and efforts were made to minimize their impact on the robustness and validity of the study's findings.

As a result, the initial sample of 288 companies was significantly reduced to 107 companies, which had available ESG scores in Bloomberg. Specifically, the final sample of 107 companies is distributed as follows:

- Metallurgy and Metal Products: 25 companies
- Chemicals, Pharmaceuticals, Petroleum, Rubber, and Plastic Products: 58 companies
- Textile and Apparel: 10 companies
- Food and Tobacco: 14 companies

This reduction in the sample was an unavoidable outcome, but given that each company spans five fiscal years, the sample size remains appropriate for this type of analysis.

The final sample used for the empirical analysis consists of 107 companies, with 535 observations in total.

MEASURES AND VARIABLES

This section provides an overview of all the variables used in the empirical analysis. The data for these variables were last verified on January 30, 2023.

The section is structured as follows: first, a description of the dependent variable (cost of debt); second, a discussion of the independent variable (ESG factors); and finally, a review of the control variables.

Dependent Variable

To calculate the cost of debt, the formula used divides "Financial Expenses" (from the income statement) by "Financial Debt" (from the balance sheet liabilities) for each company and each year. Both data points were extracted from the Orbis database. This approach was chosen to avoid methodological inconsistencies that can arise from relying on pre-calculated values, which may vary across different databases. Therefore, the cost of debt was specifically constructed for this analysis to ensure consistency and accuracy.

Orbis defines "Financial Expenses" as "financial costs, including interest expenses." To calculate "Financial Debt," we aggregated "Non-Current Liabilities" and "Debts" from the company balance sheets. According to Orbis, "Non-Current Liabilities" are defined as "long-term financial obligations, including long-term financial debt and other long-term liabilities," and "Debts" refer to "short-term financial obligations, such as short-term debt to banks and portions of long-term financial debt payable within the year."

Since financial expenses represent a flow variable, while financial debt is a stock variable, it was necessary to adjust the denominator to ensure a more accurate result. Thus, the average of financial debt at the end of year $t-1$ and year t was used in the denominator. This averaging of beginning and end-of-year debt balances from the balance sheet creates a more reliable measure. The formula for calculating the cost of debt is summarized as follows:

$$CoD = \text{Financial Expenses/Average Financial Debt}$$

The cost of debt, calculated in this manner, serves as a key financial performance indicator (KPI) that expresses, as a percentage, the impact of interest expenses on a company's total financial debt. In other words, it measures the burden of interest payments on the company's financial obligations.

The decision to use the cost of debt as the dependent variable stems from its direct relevance in assessing the financial implications of ESG practices. Unlike other financial indicators, the cost of debt explicitly reflects how financial institutions perceive a company's risk profile. ESG-aligned firms are often viewed as more stable and less exposed to environmental, social, and governance-related risks, which can lead to lower borrowing costs. By analyzing the cost of debt, we aim to capture the potential financial benefits that arise when firms integrate ESG principles into their operations.

Additionally, the cost of debt has a direct impact on a company's capital structure and overall financial performance, making it a critical variable for understanding the economic value of ESG practices. Lower borrowing costs can improve a firm's competitiveness and profitability, while higher costs may signal perceived risks or inefficiencies. Unlike equity-based measures, which can be influenced by market dynamics and investor sentiment, the cost of debt reflects a more objective assessment by creditors, who base their evaluations on tangible metrics such as creditworthiness and risk exposure.

This focus aligns with the study's objective to investigate whether ESG performance translates into tangible financial advantages, particularly in reducing the cost of borrowing. Therefore, the cost of debt serves as a robust and meaningful indicator for analyzing the financial implications of sustainability practices, providing insights into how banks and other financial institutions incorporate ESG considerations into their credit assessments.

Independent Variable

The ESG scores serve as the main variable of interest in the analysis presented in this chapter.

These scores were sourced from Bloomberg Professional System, a comprehensive database that provides financial and economic data. The Bloomberg Terminal, the flagship product of Bloomberg LP, is widely recognized as a global leader in financial information and is used by investors, analysts, consultants, and other financial professionals. It allows for real-time market analysis and research. In terms of ESG data, the Bloomberg Terminal provides both an overall score, known as the "Bloomberg ESG Score," and individual scores for each pillar of sustainability: the "BESG Environmental Pillar Score," the "BESG Social Pillar Score," and the "BESG Governance Pillar Score." These scores were downloaded for the sample companies and are proprietary to Bloomberg.

The scores range from 0 to 10, where 10 represents the highest level of sustainability performance. Bloomberg specifies that the environmental score (E_Score) assesses a company's overall environmental performance; the social score (S_Score) evaluates its social performance; and the governance score (G_Score) reflects the company's governance practices. Each of these scores encompasses multiple sub-themes that contribute to the aggregate score. In summary:

The Environmental (E) pillar includes key topics such as air quality, climate exposure, ecological impact, energy management, supply chain environmental management, greenhouse gas emissions, sustainable products, waste management, and water management.

The Social (S) pillar covers community rights and relations, corporate ethics and compliance, labor policies, occupational health and safety, operational risk management, product quality management, and social supply chain management.

The Governance (G) pillar addresses board composition (role, diversity, independence, and development), executive compensation (incentive structure and pay-for-performance), and shareholder rights (policies and voting).

The Environmental, Social, and Governance pillar scores are weighted averages of underlying theme scores. Without delving into the complex calculation methodologies (refer to the original source for details), Bloomberg's ESG score offers a consolidated measure of corporate sustainability data, simplifying the integration of ESG factors into business and investment analyses.

Regarding data collection, Bloomberg explains that its ESG scores combine various data sources, including corporate sustainability disclosures and financial fundamentals, alongside proprietary research and analytical resources (e.g., Bloomberg Intelligence's ESG research and the Bloomberg Industry Classification Standard, BICS). As a result, Bloomberg's ESG scores are grounded in evidence, rigorous research, and analytical precision. The scores are based on financially relevant data, quantitative criteria, and ensure full transparency in both methodology and the underlying data.

Control Variables

To properly test the research hypothesis and enhance the robustness of the regression model, a set of control variables was introduced, as recommended by previous literature (Hoepner et al., 2016). These control variables were selected for their relevance in influencing the CoD and their potential impact on a company's risk profile. This influence, in turn, could affect the cost of debt charged by banks. When determining the interest rate for loans, banks must consider several important factors, including the inherent risk of the loan. Risk plays a critical role in all phases of the lending process, from risk assessment (credit evaluation and rating assignment) to pricing and periodic review of the loan's profitability in terms of risk and return.

As emphasized in the literature, risk is shaped by a company's financial structure, profitability, and the specific risk level of each firm, alongside its market valuation. Other factors, such as the company's sector, geographic area, and size, can also impact risk and, consequently, the cost of debt. Therefore, financial data on control variables were collected from the Orbis database for each company and each year (2017–2021). The selected control variables are consistent with existing literature on the cost of debt, and the expected relationships between the dependent variable (CoD) and each independent variable are outlined below:

Profitability Index: Expressed by ROCE (%), which measures the return on capital employed by the company. ROCE evaluates profitability and the efficiency of capital usage by showing the profits generated for each additional unit of capital employed. Higher ROCE values typically indicate greater profitability, making these companies more attractive to investors. ROCE is expected to be negatively correlated with the cost of debt, as more profitable companies generally face lower borrowing costs.

Company Size: Represented by total assets (TA) from the balance sheet, which indicates the investments made by the company using its financial resources. Larger companies often have easier access to external financing, lower information asymmetry, and greater resilience to negative events, leading to economies of scale, even in borrowing costs. Therefore, company size is expected to be negatively correlated with the cost of debt.

Financial Leverage: Indicated by Gearing (%), which measures the ratio of debt to equity in a company. This variable represents the company's financial independence from third parties. Higher gearing ratios indicate a larger proportion of debt compared to equity, suggesting higher financial risk. Gearing is expected to be positively correlated with the cost of debt.

Market Capitalization: Refers to the total value of a publicly traded company's outstanding shares. Larger market capitalization is associated with greater stability and lower investment risk. Orbis calculates market capitalization as the number of outstanding shares multiplied by the share price. It is anticipated that higher market capitalization will result in a lower cost of debt due to the market's recognition of the company's value.

Company Risk: Represented by the company's Beta coefficient, which measures the volatility of a stock's returns relative to the market. Beta captures overall risk, including cyclical revenue risk, operating risk, and financial risk. A higher Beta indicates greater volatility, leading to increased borrowing costs. Beta is expected to be positively correlated with the cost of debt.

Industry Type: The dataset includes two capital-intensive sectors ("Metallurgy and Metal Products" and "Chemicals, Pharmaceuticals, Petroleum, Rubber, and Plastics") and two less capital-intensive sectors ("Textiles and Apparel" and "Food and Tobacco"). Industry type is included as a moderating factor to explore whether the relationship between ESG factors and the cost of debt is influenced by capital intensity.

To achieve a more normal distribution, the natural logarithm was applied to some variables, including market capitalization (MKT CAP), total assets (TA), and Gearing (%). Although not explicitly stated for brevity, all financial data in the analysis are expressed in millions of dollars, as reported in the Orbis database.

EMPIRICAL ANALYSIS

The impact of ESG factors on a company's cost of debt was examined to test the research hypothesis through the following linear regression model:

$$CoD\,t = f(ESG\,t - 1, \text{controls variables}\,t)$$

CoDt" corresponds to the cost of debt at time t, while "ESG$t-1$" refers to ESG factors from the previous year, $t-1$. All control variables included in the regression model are measured at time t, as is the dependent variable (cost of debt), since, according to the efficient market hypothesis, their impact occurs in the same year that the cost of debt is applied.

More specifically, in examining the relationship between a company's ESG factors and the cost of debt applied by banks, the following linear regression was estimated to identify the factors influencing the dependent variable (CoD):

$$i\,t = \alpha + \beta1\text{ESG}t - 1 + \beta2\text{BETA}t + \beta3\text{MKTCAP}t + \beta4\text{ROCE}t$$
$$+ \beta5\text{TA}t + \beta6\text{GEARING}t + \beta7\text{SECTOR}$$
$$+ \beta8\text{SECTOR} * \text{ESG}t - 1 + \varepsilon$$

where:

CoD = cost of debt score incurred by a company in the sample at time t, calculated as previously described;

ESG = sustainability score (ESG factors) of a company at time $t - 1$, as outlined earlier;

Beta, MKT CAP, ROCE, Total Assets (TA), and Gearing = control variables explained earlier, measured at time t;

SECTOR*ESG = interaction factor to test the moderating effect of the sector on the relationship between ESG and cost of debt;

β = effect estimate, or regression coefficient;

α = intercept, the point where the regression line crosses the y-axis;

ε = error term.

RESULTS

The regression analysis to test the hypothesis was conducted cross-sectionally on 570 observations, covering the 107 companies in the sample described above over five fiscal years (2017–2021).

It is important to note that for some companies in the sample, certain financial data were not available in the databases consulted. Consequently, there are missing data for some variables, as shown in the descriptive table below. Specifically, 16% of Beta values, 4% of Market Capitalization, and 3% of Gearing data are missing. However, it is crucial to highlight that

Table 3.1 Descriptive statistics

	N	Media	SD	Min	Max
CoD	570	3.456	7.587	−49.564	76.543
ESG_tot	570	3.675	1.465	0.000	7.567
ESG_E	570	2.856	1.871	0.000	7.654
ESG_S	570	2.454	1.234	0.000	8.543
ESG_G	570	5.567	1.765	2.454	8.554
BETA_5years	465	0.845	0.445	0.345	1.123
MKT Cap_Log	545	8.734	1.123	3.333	12.434
ROCE	570	9.654	10.654	−41.657	70.654
TA_Log	570	15.444	1.156	10.543	19.134
Gearing_Log	533	4.154	0.767	0.654	6.235

complete data are available for the primary variables—namely, the Cost of Debt and the ESG factors for the companies in the studied sample.

The data were analyzed using Stata.

The descriptive statistics of pooled data are summarized in Table 3.1.

Additionally, for descriptive purposes, a Repeated Measures ANOVA was conducted to analyze the trend of the Cost of Debt (CoD) over time, with CoD as the dependent variable and time (variable "Year") as the independent variable. The results of this analysis indicate a significant average reduction in CoD over time, $F(4,757) = 4.14$, $p < 0.01$. This trend is illustrated in Fig. 3.1 and summarized in Table 3.2.

Since it was observed that the cost of debt decreased over time (from 2017 to 2021), the variable "Year" was included in the regression analysis as an additional control variable to account for the time effect in explaining part of the variability in the cost of debt scores. Accounting for the time factor allows for a more accurate estimation of the effects of other variables, ensuring that temporal trends are properly considered in the model.

Multicollinearity

Table 3.3 on the following page presents the correlations between all the variables included in this study.

First, the results show a significant negative correlation between the total ESG score and the Cost of Debt (CoD), although the effect size is small. While this finding aligns with the hypothesis, it is important to note

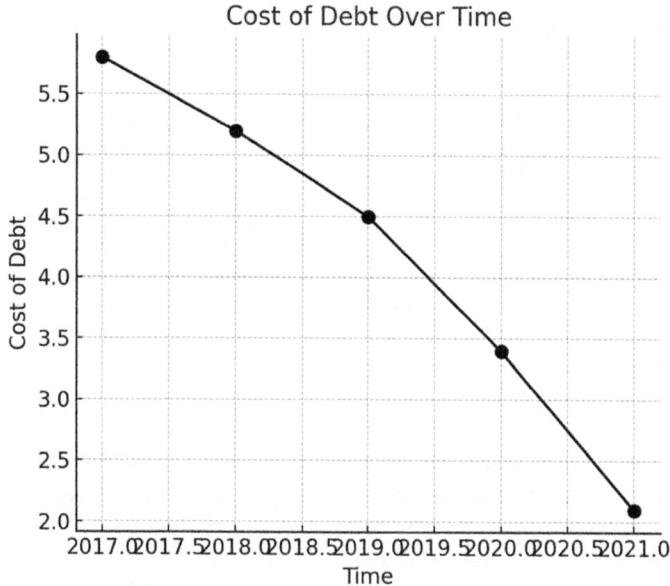

Fig. 3.1 Cost of debt over time (*Source* Own Elaboration)

Table 3.2 Descriptive statistics of CoD by year

95% confidence interval

Year	Mean	SD	Min	Max
2017	4.66	0.778	3.143	5.567
2018	4.77	0.856	3.145	6.567
2019	4.56	0.865	2.345	5.735
2020	3.22	0.558	1.134	4.764
2021	1.34	0.455	0.658	2.245

that this simple correlation analysis does not yet control for the effects of confounding variables measured in this study.

The results also indicate a positive correlation between CoD and ROCE, which is contrary to expectations. However, the prediction regarding Total Assets is confirmed, as CoD correlates negatively with

Table 3.3 Multicollinerity

	CoD	ESG_tot	ESG_E	ESG_S	ESG_G	BETA	MKT CAP	ROCE	TA	Gearing	Anno
CoD	—										
ESG_tot	0.032*	—									
ESG_E	-0.033	0.656***	—								
ESG_S	-0.012	0.787***	0.356***	—							
ESG_G	0.033	0.366***	0.166***	0.034	—						
BETA	0.033	0.273***	0.176***	0.253***	0.193***	—					
MKT CAP	-0.044	0.145***	0.224***	-0.048	0.258***	0.068	—				
ROCE	0.102*	-0.066	0.011	-0.135**	-0.024	-0.043	0.251***	—			
TA	-0.142**	0.355***	0.333***	0.220***	0.149***	0.213***	0.794***	-0.102*	—		
Gearing	-0.050	0.152***	0.021	0.154***	0.084	0.203***	-0.122*	-0.419***	0.182***	—	
Year	-0.149***	0.223***	0.244***	0.146***	0.182***	0.000	0.032	-0.091*	0.069	0.115**	—

Note $*p < 0.05$, $**p < 0.01$, $***p < 0.001$

Total Assets. Lastly, CoD is negatively correlated with Gearing, although the hypothesis had predicted a positive correlation.

Regarding the ESG factors, significant positive correlations emerge with all the control variables, except for ROCE. These significant correlation patterns underscore the importance of including Beta, ROCE, Total Assets, Gearing, and Year as control variables in the main analysis. Each variable that correlates with both ESG and CoD is a potential confounding factor.

A strong correlation ($r = 0.79$) between two predictors, Total Assets and Market Capitalization, was also observed. To avoid issues of multicollinearity in the regression analysis, only Total Assets was included as a control variable, while Market Capitalization was excluded. This decision aligns with most prior studies in the field, which focus on Total Assets rather than Market Capitalization. This choice is justified by the fact that both variables often indicate firm size and may be highly correlated.

It is important to emphasize that the simple correlations described above do not account for the effects of control variables. Therefore, a linear regression analysis is conducted to estimate the relationship between ESG factors and CoD, while controlling for potential confounders.

Among the ESG scores, only the total ESG score shows a significant correlation with CoD. As a result, the main analysis—linear regression—proceeds using only the total ESG score, rather than the individual ESG pillar scores, to explore and estimate this effect while controlling for other variables.

Analysis

Below, Table 3.4 presents the results of the fixed-effects linear regression analysis.

The overall model, considering all predictors, explains 6% of the variance in Cost of Debt values, $F(8, 429) = 3.51$, $p < 0.001$.

The results indicate a significant negative relationship between Total Assets and Cost of Debt ($\beta = -0.12$) and between Year and Cost of Debt ($\beta = -0.11$). Conversely, there is a positive relationship between ROCE and Cost of Debt ($\beta = 0.17$). However, all significant effects observed are of small magnitude ($\beta < 0.30$).

Importantly, no significant effect of ESG on Cost of Debt was found after controlling for these variables. This suggests that the previously

Table 3.4 Linear regression coefficients (dependent variable: CoD)

Variable	Coeff	SE	t	p	Std est
constant	12.6843	4.4411	2.8556	0.0042	
ESG_tot	0.0333	0.39324	0.0865	0.9351	0.00532
BETA	0.5965	1.3043	0.4556	0.6482	0.02323
TA	−0.6819	0.2934	−2.3434	0.0193	−0.12333
ROCE	0.1444	0.0435	3.2878	0.0015	0.17112
Gearing	0.3376	0.5016	0.6610	0.5054	0.03656
Year	−0.6278	0.2754	−2.2234	0.0233	−0.11122
BvDsectors	1.11243	2.6467	0.4223	0.6752	−0.08023
BvDsectors * ESG_tot:	−0.5014	0.8296	−0.6143	0.5383	−0.07932

observed significant negative correlation between CoD and ESG was spurious, and other factors (such as Total Assets, Time, and ROCE) play a more substantial role in predicting variations in the Cost of Debt.

Additionally, no significant interaction effect between "Sector" and ESG was observed, indicating that the non-significant effect of ESG applies across both capital-intensive and less capital-intensive sectors, as outlined in the sample description.

Regarding the Year variable, the results show that, on average, the Cost of Debt decreases by 0.62 points each year (a decrease of 0.11 SD).

For Total Assets, the findings indicate that for each million-dollar increase in total assets during a given year, the Cost of Debt decreases by 0.68 points (a decrease of 0.12 SD).

As for ROCE, the results demonstrate that for every percentage point increase in Return on Capital Employed in a given year, the Cost of Debt increases by 0.14 points (an increase of 0.17 SD).

Discussion

This study investigates the relationship between the cost of debt (CoD) and companies' ESG factors, aiming to assess whether firms committed to sustainability benefit from improved loan conditions, specifically a lower CoD. Our hypothesis posited a negative correlation between ESG factors and CoD, suggesting that higher ESG scores could lead to lower borrowing costs as a reward from banks, reflecting the integration of ESG factors into lending evaluations (Bauer & Hann, 2010; Hoepner et al., 2016). Although initial correlation analysis showed a significant negative relationship between ESG and CoD, this significance disappeared in the regression analysis after controlling for other variables. This suggests that other factors, such as Time (Year), Return on Capital Employed (ROCE), and Total Assets, play a more prominent role in explaining CoD variability over the period and within the sample considered, indicating that our research hypothesis cannot be confirmed for this dataset.

Among the factors significantly affecting CoD, Time (Year) showed that CoD declined over the study period (2017–2021). This period includes the COVID-19 pandemic, which led the European Central Bank (ECB) to implement policies aimed at mitigating economic impacts by keeping interest rates low, affecting bank lending rates and facilitating access to credit for firms (Kohlscheen, 2021). These monetary policy adjustments may explain the observed decline in CoD, highlighting how external economic interventions can overshadow firm-specific ESG impacts during extraordinary periods. Additionally, our findings showed a counterintuitive positive relationship between ROCE and CoD, contradicting prior literature suggesting that higher profitability should correlate with lower CoD (Goss & Roberts, 2011; Hasan et al., 2022). Future studies might consider alternative profitability indicators, such as ROI or ROA, to further investigate this anomaly. Consistent with previous findings, firm size, proxied by Total Assets, demonstrated a negative correlation with CoD, suggesting that larger firms benefit from lower borrowing costs due to perceived stability and lower risk profiles (Ge & Liu, 2015).

The absence of a significant effect for ESG factors suggests that ESG considerations alone do not strongly influence CoD. This finding aligns with prior studies showing inconsistent results in the ESG-CoD relationship, likely due to market-specific factors, temporal context, and sectoral variations (Devalle et al., 2017; Nandy & Lodh, 2012). This study also highlights possible sampling limitations: despite selecting publicly traded firms to reduce missing data, ESG scores were available for only a limited number, reducing the sample size. Moreover, our focus on specific sectors—metallurgy, chemicals, textiles, and food—may have influenced results. These sectors have faced stringent EU sustainability regulations only recently, and the potential long-term effects of ESG adoption on debt costs may not yet be evident.

Finally, our findings underscore the challenges of drawing consistent conclusions about ESG's impact on CoD across sectors. While no significant sectoral interaction effect with ESG was observed, future research could test for potential moderating effects with a larger sample per sector. Our study contributes to existing literature by focusing on specific EU industries relevant to the sustainable transition and considering a unique timeframe that includes the unprecedented COVID-19 period. In line with existing research, these findings underscore the complex, often inconsistent relationship between ESG factors and debt financing costs, particularly when traditional financial determinants remain influential.

CONCLUSION

The empirical findings of this study indicate that there is no clear and significant relationship between the cost of debt (CoD) and ESG factors within the sample analyzed. Importantly, the results suggest a neutrality in how ESG factors are considered, with no positive or negative bias detected in lending rates based on ESG performance. This neutrality implies that while ESG factors have yet to become a decisive criterion for banks, they are also not perceived as adding risk, suggesting that credit decisions still rely more heavily on traditional financial metrics rather than sustainability criteria (Ge & Liu, 2015; Goss & Roberts, 2011).

The lack of significant results aligns with existing literature that finds mixed or inconclusive links between ESG and debt costs, highlighting the complex, evolving nature of sustainable finance and its integration into lending practices (Devalle et al., 2017; Hoepner et al., 2016). This study contributes to this body of research, encouraging future investigations to assess whether specific components of ESG, such as governance quality or environmental performance, hold greater weight in credit decisions across particular sectors or regions. Disentangling the three ESG pillars—environmental, social, and governance—could provide more nuanced insights into their respective impacts on CoD, as different pillars may play distinct roles depending on sectoral and regional characteristics. Future research could build on this by analyzing these pillars individually, exploring whether specific ESG dimensions are more influential in credit risk assessments.

Indeed, recent regulatory efforts by the European Union, such as the EU Taxonomy for Sustainable Finance and the Sustainable Finance Disclosure Regulation (SFDR), are pushing financial institutions to account more thoroughly for ESG factors in their assessments, suggesting that ESG-related risk and value might gradually become more influential in determining CoD (European Commission, 2021).

It is also necessary to recognize the limitations of this analysis, particularly regarding timing and effect sizes. ESG integration in financial systems is a relatively recent phenomenon, and banks, like other market actors, may require time to meaningfully adjust their models and incorporate sustainability factors into credit assessments (Schoenmaker & Schramade, 2019). The regulatory environment itself is still evolving, with initiatives like the CSRD and SFDR progressively being implemented. These regulations are expected to enhance transparency, improve data quality, and standardize ESG metrics, which could clarify the connection between ESG practices and the cost of debt (CoD) over time (Pizzi et al., 2021).

In addition to timing challenges, the modest effect sizes for certain control variables, such as Total Assets and Year, should also be acknowledged as a limitation. While statistically significant, these small effect sizes suggest that their influence on CoD may be limited, restricting the generalizability of broader conclusions about their impact.

Furthermore, the sample size and data constraints in this study may have affected the findings. With only 107 companies possessing complete ESG scores and a limited timeframe from 2017 to 2021, the analysis may not have fully captured the long-term impact of ESG efforts, especially given that sustainability initiatives often require considerable time to influence financial outcomes. The analysis period also coincides with the COVID-19 pandemic, which dominated the final years of the study and could have disrupted emerging ESG-finance trends, as financial institutions prioritized immediate recovery efforts over long-term sustainability considerations (Kohlscheen, 2021; Schoenmaker, 2017). These limitations underscore the need for continued research to explore the evolving dynamics between ESG practices and financial performance over longer periods and across larger datasets.

Nonetheless, the growing alignment among EU regulators, financial institutions, and companies signals an increasing prioritization of sustainability in financial practices. As ESG reporting standards improve and regulatory pressures strengthen, it is likely that ESG considerations will become more central to credit risk assessments. Future research should therefore seek to capture this dynamic by incorporating longer-term data, industry-specific analysis, and the evolving regulatory landscape. Additionally, exploring the individual effects of the three ESG pillars could provide critical insights into how each dimension influences CoD under varying conditions.

In conclusion, while this study did not observe a significant impact of ESG factors on CoD, it lays the groundwork for further inquiry, underscoring the need for more granular data and robust frameworks that can reflect the nuanced role of ESG in credit markets. Future studies should consider disentangling the three ESG pillars to better understand their specific roles in shaping financial decisions, particularly as the transition to a low-carbon economy promises to reshape corporate finance and risk management.

CHAPTER 4

Future Directions and Innovations in Sustainable Finance: Challenges and Opportunities

Abstract This chapter explores the transformative role of green fintech and digital transformation in advancing sustainable finance. Green fintech, which integrates financial technology with sustainability goals, leverages innovations like blockchain, artificial intelligence (AI), and digital platforms to enhance ESG (Environmental, Social, and Governance) practices. Blockchain improves transparency and accountability, particularly in supply chain traceability and carbon credit trading, while AI refines ESG data analysis and risk assessments. Digital platforms democratize sustainable finance, enabling broader access to green investments through tools like crowdfunding and micro-investing. Despite its potential, green fintech faces challenges, including data privacy, cybersecurity, and regulatory compliance, necessitating international cooperation to standardize frameworks. The chapter concludes that green fintech is pivotal in bridging digital finance with sustainability, fostering a resilient financial ecosystem that prioritizes environmental and social advancements.

Keywords Green fintech · Sustainable finance · Blockchain technology · Artificial intelligence (AI) · ESG (Environmental Social and Governance) · Digital transformation

© The Author(s), under exclusive license to Springer Nature Switzerland AG 2025
N. Del Sarto, *ESG Factors and Financial Outcomes in Banks*,
https://doi.org/10.1007/978-3-031-87748-3_4

GREEN FINTECH AND DIGITAL
TRANSFORMATION IN SUSTAINABLE FINANCE

The rapid evolution of financial technology, or "fintech," is reshaping the sustainability landscape by providing new tools for enhancing transparency, accessibility, and accountability in sustainable finance. Green fintech, a term that encapsulates the intersection of technology and environmentally-oriented financial solutions, enables more efficient management of ESG (Environmental, Social, and Governance) factors and promotes sustainability across the financial industry. Fintech innovations like blockchain, artificial intelligence (AI), mobile platforms, and digital assets are now central to advancing green finance by facilitating the tracking of sustainable practices, lowering costs, and broadening access to green financial products for investors and consumers alike (Chen et al., 2020).

One of the most transformative aspects of green fintech is blockchain technology. Blockchain, a decentralized and secure ledger system, enhances the traceability and transparency of supply chains, a crucial factor in monitoring the authenticity of ESG commitments. By ensuring data immutability and enabling real-time tracking, blockchain allows companies and investors to verify claims about sustainability practices, such as carbon footprint reductions or ethical sourcing of materials (Wang et al., 2021). Blockchain applications can help counteract issues like greenwashing by providing a verifiable, public record of ESG practices, thereby increasing trust among stakeholders (Tang & Demeritt, 2020). Additionally, blockchain's decentralized structure can facilitate cross-border investments in green bonds and projects, which is essential for channeling global capital into developing regions where funding is often scarce but environmental impact potential is significant. Furthermore, green asset-backed tokens and blockchain-enabled carbon credit trading platforms allow for the direct and efficient exchange of value tied to environmental initiatives, increasing accountability in carbon offset programs (Lee et al., 2022).

Digital currencies and platforms also play a pivotal role in democratizing sustainable finance, making it more accessible to a broader audience. Mobile apps and online platforms enable investors of all sizes to participate in green investments, whether through crowdfunding for renewable energy projects or purchasing shares in green bonds. These tools not only lower entry barriers but also allow for micro-investing in sustainable products, a shift that aligns with the growing consumer demand for accessible and responsible investment options (Zhou & Tian, 2019). This democratization of sustainable finance has particular significance in promoting social equity. For instance, crowdfunding platforms enable small-scale investors and marginalized communities to support local sustainability projects, which can enhance local environmental quality and create community-driven growth opportunities (Boreiko & Massarotto, 2021). This accessibility is essential in enabling the public to contribute to sustainability, fostering inclusivity and empowering communities to drive environmental progress.

Artificial intelligence (AI) further enhances sustainable finance by refining ESG data analysis and risk assessment. AI algorithms can analyze massive datasets to provide insights into ESG risks, such as the potential impacts of climate change on specific industries or supply chains. This analysis helps investors make informed decisions based on comprehensive risk factors rather than just financial returns, shifting the focus toward long-term sustainability. For example, AI-powered tools can assess companies' carbon footprints or analyze social factors, such as labor practices, by pulling from sources like social media, government reports, and satellite imagery (Ding & Li, 2020). In addition, AI applications in fintech facilitate ESG scoring and benchmarking by providing real-time data and predictive insights into a company's environmental and social performance, crucial elements for investors increasingly attuned to sustainability (Yang & Li, 2021). Through machine learning models, AI can also predict the future impact of environmental policies or resource depletion on financial returns, thereby enabling more resilient, future-proof investment strategies. Additionally, AI in predictive climate modeling helps companies prepare for climate-related risks, ensuring that portfolios are aligned with both profit and planetary health.

Despite its transformative potential, the integration of fintech into sustainable finance is not without challenges. Data privacy, cybersecurity, and regulatory compliance remain pressing concerns. Given that ESG data often includes sensitive company information, securing this data against breaches and unauthorized access is essential for maintaining trust. Furthermore, as financial regulators seek to standardize ESG disclosures, fintech firms and financial institutions must adapt to comply with evolving legal frameworks, adding a layer of complexity to the deployment of green fintech solutions (Schoenmaker & Schramade, 2019). Additionally, the decentralized nature of blockchain, while enhancing transparency, raises new regulatory challenges, as traditional financial regulations are not always suited to decentralized digital assets. This regulatory gap underscores the need for international cooperation and the development of specific standards to govern the use of blockchain and digital finance in sustainability contexts (Zysman & Huberty, 2014).

In conclusion, green fintech is poised to play a critical role in the global push towards a sustainable economy. By enhancing transparency, accessibility, and accountability, digital tools and platforms are helping to reshape the financial sector's approach to sustainability. As technological innovation continues, green fintech could serve as a catalyst for broad-based sustainable economic transformation, bridging the gap between digital finance and responsible, inclusive growth. By strategically employing digital innovations such as blockchain, AI, and mobile finance, green fintech has the potential to create a more resilient financial ecosystem where financial gains go hand-in-hand with environmental and social advancements. As the demand for sustainable finance rises, these technologies could not only meet market needs but also contribute to a new global standard for economic growth that emphasizes both sustainability and inclusivity.

IMPACT INVESTING AND MEASURING SOCIAL AND ENVIRONMENTAL RETURNS

In recent years, impact investing has emerged as a transformative approach within the broader field of sustainable finance, distinguishing itself by its dual focus on financial returns and measurable social and environmental outcomes. While traditional investments primarily aim for profit, impact investing aligns financial objectives with intentional, positive societal and environmental impacts. This section explores the key aspects of impact investing, from defining its goals to discussing the methodologies and metrics used for assessing outcomes. Additionally, it highlights the evolving expectations of investors and how impact investing serves the wider objectives of sustainable finance.

Defining Impact Investing

Impact investing, often considered a subset of sustainable finance, goes beyond traditional investing by actively pursuing positive change. Defined by the Global Impact Investing Network (GIIN), impact investing seeks to generate measurable social and environmental impact alongside financial returns (GIIN, 2021). This investment strategy represents a departure from conventional models by prioritizing outcomes that support societal well-being, address global environmental challenges, and advance economic inclusivity (Brest & Born, 2013). Impact investments span a wide array of sectors, including clean energy, affordable housing, healthcare, and education, where they tackle issues such as climate change, poverty, and access to quality services (Bugg-Levine & Emerson, 2011).

A fundamental characteristic of impact investing is **intentionality**—the explicit intention to drive positive change. This sets it apart from other sustainable investments where the impact may be secondary or incidental to financial goals (Clark et al., 2015). Additionally, impact investors are increasingly committed to **accountability**, requiring rigorous measurement and reporting to ensure that investments achieve their desired social and environmental effects.

Methodologies and Metrics for Measuring Impact

A key challenge in impact investing lies in accurately measuring and verifying social and environmental outcomes. Effective impact measurement not only allows investors to track the progress of their investments but also helps build credibility and transparency in the impact investing sector. To address this, various methodologies and frameworks have been developed, focusing on standardized metrics, reporting guidelines, and performance evaluation.

1. **Impact Reporting and Investment Standards (IRIS)**: Developed by the GIIN, IRIS provides a catalog of standardized metrics for measuring impact across multiple sectors. These metrics allow investors to evaluate specific areas such as job creation, environmental conservation, and health improvements, providing comparability across investments (Jackson, 2013).

2. **Global Reporting Initiative (GRI)**: Although originally designed for corporate sustainability reporting, GRI standards are widely used by impact investors to assess social, environmental, and economic performance. The GRI framework emphasizes stakeholder engagement, materiality, and accountability, offering a comprehensive view of an investment's impact on stakeholders (GRI, 2011).

3. **Theory of Change:** This methodology outlines how specific activities lead to desired outcomes, allowing investors to establish a roadmap of anticipated impact. By breaking down the causal relationships between actions and outcomes, the Theory of Change helps identify key impact indicators and assess whether interventions are effective (Weiss, 1995).
4. **Sustainable Development Goals (SDGs):** The United Nations' SDGs provide a global framework for addressing pressing social and environmental issues. Investors increasingly use the SDGs to align impact goals, with specific metrics focused on poverty alleviation, clean energy, health, and education. This alignment enables impact investors to contextualize their goals within a broader, globally recognized framework (UN, 2015).

The choice of methodology depends on the investor's goals and sectoral focus. For example, IRIS metrics might be ideal for measuring impact in microfinance or renewable energy, while Theory of Change might better suit interventions in complex social issues like healthcare or education. By standardizing metrics and embracing rigorous methodologies, impact investors are enhancing their ability to track progress and communicate results transparently.

The Evolving Expectations of Investors

Investor expectations in the realm of impact investing have evolved significantly, driven by a growing recognition that sustainable economic growth requires solutions that address systemic issues. Traditionally, investors balanced financial performance with basic compliance or philanthropic efforts. However, the increasing focus on climate change, social equity, and corporate accountability has heightened expectations for meaningful and measurable impact, moving beyond mere risk mitigation to active contributions toward global well-being (Eccles & Klimenko, 2019).

The demand for transparency has also grown, with investors now expecting robust data on the social and environmental outcomes of their investments. To meet these expectations, organizations are adopting **impact-weighted accounts**, a system that monetizes the impact of a company's activities, including its environmental footprint, social contributions, and governance practices (Serafeim, 2020). Such measures allow investors to directly compare financial performance with non-financial outcomes, creating a more integrated view of overall impact.

Moreover, the rise of **millennial and Gen Z investors**, who prioritize purpose alongside profit, has catalyzed demand for impact investing products that align with personal values. According to a Morgan Stanley study (2019), 85% of millennials are interested in sustainable investing, suggesting that future generations are likely to continue shaping the field with heightened expectations for measurable outcomes and transparent reporting.

The Role of Impact Investing in Sustainable Finance

Impact investing is a cornerstone of sustainable finance, not only fulfilling investors' goals for responsible returns but also contributing to the broader objectives of sustainability. By channeling capital into socially and environmentally beneficial projects, impact investing mobilizes resources toward addressing global challenges in a way that traditional philanthropy and market-based investments alone cannot achieve (Bugg-Levine & Emerson, 2011). As a bridge between profit-driven and purpose-driven finance, impact investing aligns with the sustainability goals of governments, businesses, and individuals, reflecting an ecosystem approach that combines financial returns with societal benefits.

Impact investing also contributes to **market innovation** by encouraging the development of sustainable solutions across industries. For instance, investments in renewable energy have accelerated technological advancements in solar, wind, and bioenergy, while impact capital in healthcare has improved access to affordable services in underserved regions. This type of capital allocation not only supports individual businesses but also stimulates innovation, enhances sector resilience, and supports a low-carbon economy (Schoenmaker & Schramade, 2019).

In addition, impact investing addresses systemic social issues by focusing on underserved or marginalized communities. Financial inclusion initiatives, for example, have increased access to financial services in low-income areas, promoting economic empowerment and reducing poverty (Banerjee & Jackson, 2017). By promoting equitable growth, impact investing contributes to sustainable economic resilience, fostering a society in which both businesses and individuals can thrive.

Challenges and the Future of Impact Investing

While impact investing holds significant potential, challenges remain, particularly concerning standardization, scalability, and verification. Impact metrics and reporting, though advanced, are not yet universally standardized, leading to inconsistent measurement practices and potential issues with greenwashing or impact dilution (Bowman, 2017). Investors and regulatory bodies must address these concerns to maintain the sector's credibility and to ensure that impact investing achieves its stated objectives.

Moreover, as impact investing grows, the scalability of projects remains a key challenge. Large institutional investors, such as pension funds and sovereign wealth funds, often require projects with substantial capital and clear, quantifiable returns, which can be difficult to achieve in nascent or underserved sectors (Martin & Gregory, 2015). Addressing scalability will require innovative financing structures, public–private partnerships, and increased government support to create an ecosystem where impact investments can grow and thrive.

Despite these challenges, the future of impact investing looks promising, with continued advancements in data analytics, impact measurement, and market structures. As technology evolves, it will further enable impact investors to analyze complex datasets, track real-time impact, and create adaptive strategies for maximizing positive outcomes. These advancements will ultimately enhance the capacity of impact investing to contribute to sustainable development, reinforcing the role of finance in driving social and environmental progress.

Impact investing has redefined the boundaries of finance by integrating financial performance with intentional, measurable social and environmental outcomes. As investors increasingly demand accountability and transparency, the methodologies and metrics for assessing impact continue to evolve, laying a strong foundation for future growth. Impact investing not only addresses immediate societal needs but also supports long-term sustainability goals, playing a crucial role in aligning the financial sector with broader environmental and social objectives. With growing interest from individuals and institutions alike, impact investing is poised to shape the future of sustainable finance, advancing a more resilient, equitable, and sustainable economy.

The Role of Artificial Intelligence and Big Data in ESG Analysis

As sustainable finance increasingly gains traction, the demand for accurate, timely, and comprehensive ESG (Environmental, Social, and Governance) data has become crucial. Artificial Intelligence (AI) and big data analytics are emerging as transformative tools in this field, offering new ways to collect, analyze, and interpret ESG data. From predictive models that help evaluate environmental risks to sentiment analysis that assesses social and governance factors, AI and big data are redefining how financial markets and institutions incorporate sustainability metrics. This chapter explores the contributions of AI and big data to ESG analysis, examining both the opportunities and ethical challenges inherent in leveraging these advanced technologies within sustainable finance.

AI and big data analytics bring substantial advantages to ESG analysis, particularly by addressing the often fragmented and unstructured nature of sustainability data. Traditional ESG assessment methods rely heavily on corporate disclosures, reports, and surveys, which are often limited in scope, inconsistent, or outdated by the time they reach analysts. AI and big data, however, can process massive volumes of diverse data sources—such as satellite imagery, social media content, news reports, and environmental sensors—enabling real-time insights that are far more granular and dynamic than previously possible (Eccles et al., 2014). This data-driven approach provides a more comprehensive understanding of companies' environmental impacts, social policies, and governance structures.

For instance, AI-powered predictive analytics can analyze climate patterns and predict environmental risks, such as floods, wildfires, or droughts, affecting corporate assets. These capabilities are essential for investors and asset managers aiming to understand and mitigate climate-related financial risks (Bolton et al., 2020). Similarly, AI enables sentiment analysis on social media and news platforms to gauge public opinion, employee satisfaction, and community relations, offering valuable insights into a company's social and governance practices (Cade, 2018). By automating data collection and analysis, AI and big data allow stakeholders to evaluate ESG performance more accurately and holistically, aligning investment strategies with sustainability goals.

One of the most significant applications of AI and big data in ESG analysis is in enhancing predictive capabilities for environmental risk assessment. Leveraging machine learning algorithms and large datasets, AI can predict potential environmental risks by identifying historical patterns and correlations among environmental variables. For example, AI can track deforestation rates, water usage, and pollution levels, predicting how these factors may evolve and impact ecosystems and communities (Caldecott et al., 2015). By analyzing satellite images, weather data, and other environmental records, AI systems can offer investors detailed risk assessments that help them understand and avoid unsustainable investments (Schoenmaker & Schramade, 2019).

For companies, predictive analytics help improve resilience by identifying ESG-related risks within their supply chains or operational sites. This capability is particularly valuable in sectors with high environmental exposure, such as agriculture, energy, and manufacturing. By evaluating potential risks, companies can make strategic adjustments that mitigate their environmental footprint and reduce financial losses from ESG-related incidents.

Sentiment analysis, a form of natural language processing (NLP), is another critical application of AI in ESG analysis. By scanning large volumes of textual data, such as news articles, social media posts, and customer reviews, AI algorithms can measure public sentiment toward a company and assess its social and governance performance (Bollen et al., 2011). This approach offers a more immediate understanding of how companies are perceived by employees, customers, and other stakeholders, which is particularly valuable given the increasing importance of corporate reputation in ESG scoring.

For example, if a company is embroiled in a labor dispute, sentiment analysis can track changes in public opinion and employee satisfaction over time, alerting investors to potential social risks that may affect the company's financial performance (Huang et al., 2020). Additionally, sentiment analysis provides real-time updates on governance issues such as executive controversies, legal disputes, and regulatory compliance, allowing for a dynamic evaluation of a company's adherence to ESG principles.

Despite the advantages of AI and big data in ESG analysis, their application presents several ethical and technical challenges. A primary concern is the accuracy and reliability of data sources. Since AI models rely on large, diverse datasets, the quality of ESG analysis is highly dependent on data accuracy. Data sources like social media or news content can be biased or incomplete, and AI algorithms trained on such data may yield skewed results, leading to flawed ESG assessments (Pérez et al., 2021).

Transparency in AI algorithms is also a critical issue. Many AI models operate as "black boxes," making it difficult for users to understand how they reach specific conclusions. This lack of transparency poses risks in the context of ESG analysis, where stakeholders demand clear and accountable reporting. Without algorithmic transparency, investors may find it challenging to validate ESG scores or make informed decisions, potentially undermining trust in AI-driven ESG assessments (Raji et al., 2020). To address this, researchers and practitioners advocate for explainable AI (XAI) frameworks that clarify AI decision-making processes, ensuring that stakeholders can interpret and trust AI-generated insights (Gunning et al., 2019).

Another ethical concern is bias in AI algorithms. AI systems are only as impartial as the data they are trained on, which means that biases in training data can influence ESG ratings. For example, if an AI system relies on historical data that reflects past discrimination in lending or hiring practices, it may inadvertently perpetuate these biases in its assessments. Ensuring fairness in AI-driven ESG analysis is essential to prevent discrimination against certain companies, sectors, or regions (Mehrabi et al., 2021). Addressing these biases requires developing and implementing standards for data collection, algorithm training, and performance evaluation, ultimately creating more robust, equitable AI models for ESG analysis.

The potential for AI and big data in ESG analysis is expansive, and advancements in AI technology are likely to further refine ESG assessment practices. As algorithms become more sophisticated and capable of handling complex datasets, AI-driven ESG models may integrate multifactor analysis, combining environmental, social, and governance data to provide a comprehensive risk and performance assessment. This integrated approach would offer investors a more holistic view of a company's sustainability, aligning financial returns with broader societal goals (Schoenmaker & Schramade, 2019).

Furthermore, blockchain technology has shown promise in complementing AI in ESG data management by ensuring data integrity, transparency, and traceability. By securely recording ESG metrics on decentralized ledgers, blockchain can reduce data manipulation risks and enhance accountability, making it a valuable tool for impact verification in sustainable finance (Tapscott & Tapscott, 2016).

In addition, collaborative AI platforms could foster data sharing and cooperation among stakeholders, helping standardize ESG data and metrics across industries. Collaborative platforms would enable companies, investors, and regulators to share insights and develop industry-wide standards for ESG analysis, ultimately supporting a more transparent and efficient approach to sustainability (Kraus et al., 2020).

AI and big data have revolutionized ESG analysis by enhancing data collection, risk assessment, and performance evaluation processes, bringing ESG insights to new levels of accuracy and transparency. Predictive analytics and sentiment analysis allow stakeholders to make informed decisions, integrating environmental, social, and governance factors into sustainable investment strategies. However, the application of AI in ESG analysis is not without challenges, including concerns around data accuracy, transparency, and algorithmic bias. Addressing these ethical issues is essential to fully realizing the potential of AI in sustainable finance. As AI technology continues to evolve, it will likely play an increasingly central role in advancing ESG practices, contributing to a financial ecosystem that supports both profitability and societal well-being.

Regulatory and Policy Innovations for Sustainable Finance

The drive for sustainable finance has catalyzed a series of regulatory innovations designed to enhance transparency, support green investment, and mitigate climate-related risks. As sustainable finance gains global importance, regulators and policymakers are implementing frameworks to ensure that environmental, social, and governance (ESG) considerations are fully integrated into financial practices. This chapter examines recent developments in regulatory and policy approaches, assesses the role of disclosures, and considers how various jurisdictions are shaping sustainable finance practices.

One of the most significant trends in sustainable finance regulation is the push for more rigorous and standardized ESG disclosures. Investors and stakeholders increasingly demand transparency about a company's environmental and social impacts. The European Union (EU) has led these efforts with pioneering regulations, most notably the Sustainable Finance Disclosure Regulation (SFDR), which mandates ESG disclosures by financial market participants (Schoenmaker & Schramade, 2019). The SFDR requires firms to disclose how they consider ESG risks in investment processes, thereby promoting transparency and reducing greenwashing risks (European Commission, 2020).

Additionally, the EU Taxonomy Regulation categorizes economic activities based on their contribution to sustainability objectives, including climate change mitigation and adaptation. This regulation aims to create a common understanding of what constitutes a sustainable investment, thus enabling investors to make more informed choices (European Commission, 2019). The United Kingdom and the United States are following suit with their own frameworks. The U.S. Securities and Exchange Commission (SEC), for example, is working on a rule to standardize climate-related disclosures, which could have a significant impact on how American companies report ESG data (KPMG, 2021).

Global policy efforts are increasingly focused on channeling capital towards green investments. In 2021, the EU launched the Green Bond Standard (GBS), designed to increase transparency and reliability in the green bond market. The GBS aims to enhance investor confidence by ensuring that proceeds are directed to environmentally sustainable projects (Climate Bonds Initiative, 2021). Similarly, China has made significant strides in this area, with a government-led initiative to develop green bond guidelines that align with global standards (People's Bank of China, 2020).

In the United States, tax incentives for renewable energy investments encourage capital allocation toward green projects. However, the U.S. regulatory approach differs from that of the EU and China, as it relies more on market-driven incentives rather than mandatory frameworks (Sullivan & Gouldson, 2020). This difference highlights the diversity in regulatory approaches, which affects how quickly and efficiently green investment practices are adopted worldwide.

The link between climate-related risks and financial stability is increasingly evident, with regulators developing guidelines to manage these risks within financial institutions. The Network for Greening the Financial System (NGFS), a coalition of central banks and supervisors, has spearheaded these efforts, offering climate risk assessment guidelines to support central banks in managing environmental risks (NGFS, 2019). The NGFS recommends integrating climate scenarios into stress tests to help financial institutions anticipate how climate events might impact their portfolios (Battiston et al., 2017).

The Bank of England was one of the first central banks to adopt climate stress testing, and it set a precedent that other central banks, such as those in France and the Netherlands, have since followed. In the U.S., the Federal Reserve has taken steps to study the financial impacts of climate change but has yet to implement formal stress-testing requirements. These differences underscore a varied global approach to climate-related risk assessments, with European regulators generally adopting more rigorous policies than their U.S. counterparts (Carney, 2021).

A comparative analysis of regulatory approaches across regions reveals distinct strategies for promoting sustainable finance. The EU's approach is characterized by its emphasis on mandatory disclosures, taxonomies, and incentives to foster green investments. This comprehensive strategy reflects a strong regulatory stance on sustainable finance, with the objective of positioning Europe as a leader in green finance (European Commission, 2021). China, on the other hand, has integrated sustainable finance as part of its Five-Year Plans and relies heavily on state intervention to direct funding toward green projects (Zhou & Wang, 2020).

The United States, however, adopts a more decentralized and market-driven approach. While federal-level initiatives such as the SEC's disclosure requirements are emerging, state-level actions and private sector standards are predominant. This divergence highlights a potential challenge in global coordination, as differing regulations can create barriers to cross-border sustainable investments (Sullivan & Gouldson, 2020).

Looking forward, international organizations such as the International Financial Reporting Standards (IFRS) Foundation are working towards creating a global baseline for sustainability disclosures. The formation of the International Sustainability Standards Board (ISSB) marks an effort to harmonize ESG reporting standards worldwide, which could reduce fragmentation in global sustainability reporting and facilitate cross-border investments (IFRS Foundation, 2021).

Moreover, the Financial Stability Board (FSB) is developing guidelines for integrating ESG risks into the global financial stability framework. This initiative reflects a recognition that a coordinated approach to sustainable finance is essential for managing global risks. As international cooperation intensifies, a more unified regulatory landscape is expected to emerge, enabling sustainable finance to scale globally and become a mainstream component of financial markets.

The regulatory landscape of sustainable finance is evolving rapidly, with notable strides in disclosures, green investment frameworks, and climate risk assessments. Differences in regional approaches highlight the need for international collaboration to achieve coherence in ESG standards. As sustainable finance continues to grow, these regulatory innovations will play a crucial role in fostering transparency, reducing environmental risks, and supporting the transition to a low-carbon economy. Through a unified and robust regulatory framework, sustainable finance can support a resilient and sustainable global economy, aligning financial markets with environmental and social goals.

CHALLENGES AND OPPORTUNITIES IN SCALING SUSTAINABLE FINANCE

The evolution of sustainable finance is reshaping the global economy, offering new pathways for resilience, innovation, and sustainable growth. Financial institutions, regulators, and investors are increasingly committed to sustainable practices, but they face a range of challenges and opportunities. This section delves into the major obstacles sustainable finance encounters and the key prospects for growth within this transformative sector.

Combatting Greenwashing and Enhancing Transparency

One of the most persistent challenges in sustainable finance is greenwashing—when companies misrepresent their sustainability efforts, often exaggerating or fabricating environmental and social claims. Greenwashing misleads investors and distorts the flow of capital away from authentically sustainable initiatives, undermining trust in ESG (Environmental, Social, Governance) markets (Delmas & Burbano, 2011). The European Union's Sustainable Finance Disclosure Regulation (SFDR) addresses this by requiring financial institutions to report their sustainability practices transparently and consistently (European Commission, 2020). However, regulatory gaps and the lack of harmonized global ESG standards pose ongoing challenges. Countries and jurisdictions have varied requirements, making it difficult for investors to verify whether companies meet genuine sustainability criteria.

As greenwashing has financial and reputational consequences, many financial actors advocate for the establishment of uniform ESG guidelines to ensure authenticity. Effective solutions could involve third-party verification processes, standardized ESG certifications, and incentives for companies to conduct rigorous audits of their environmental impact. The development of global ESG standards, currently pursued by initiatives like the International Sustainability Standards Board (ISSB), aims to curb greenwashing by creating more reliable, comparable frameworks for assessing sustainability. Achieving this would allow the sustainable finance sector to direct capital more effectively toward impactful projects, reinforcing its credibility as a powerful force for positive environmental and social change.

Overcoming the Lack of Standardized ESG Metrics

The absence of universally accepted ESG metrics remains a major barrier in scaling sustainable finance, as it complicates the assessment, comparability, and consistency of ESG reporting across industries and regions. Traditional financial metrics are well-established and easily compared, but ESG metrics still lack this level of standardization, with varying interpretations across rating agencies, frameworks, and regions. This inconsistency can result in confusion, misaligned priorities, and even mistrust among investors attempting to evaluate the sustainability performance of companies (Kotsantonis et al., 2016).

To address this, global standardization efforts like the IFRS Foundation's ISSB and the Task Force on Climate-related Financial Disclosures (TCFD) are developing frameworks that aim to consolidate and align disparate approaches, establishing clear, comparable standards for ESG reporting (IFRS Foundation, 2021). Such initiatives aim to enable investors to make well-informed decisions, providing a consistent basis for evaluating sustainability. Standardized ESG metrics would bring greater transparency to financial markets, increasing the credibility of sustainable investments and helping financial actors distinguish between firms with genuine sustainability commitments and those with only superficial ESG claims. As these frameworks gain traction, they hold the potential to advance sustainable finance significantly by creating a more reliable foundation for decision-making and fostering investor confidence in the ESG credentials of businesses.

Leveraging Technological Innovation for Sustainable Finance

Technological advancements, particularly in artificial intelligence (AI) and big data, offer transformative opportunities for sustainable finance, enhancing ESG analysis, data management, and risk assessment. AI-driven predictive analytics enable financial institutions to identify and anticipate environmental risks, offering insights into climate vulnerabilities and helping companies mitigate potential impacts. For example, machine learning algorithms can process vast amounts of data from various sources, improving the accuracy of ESG reporting and uncovering discrepancies in sustainability disclosures (Shiller, 2017).

Blockchain technology also offers exciting potential for sustainable finance, as it provides a secure, immutable ledger to record and verify ESG data. This transparency can increase investor confidence by ensuring the integrity of sustainability data, making it more difficult for companies to engage in greenwashing or misreporting (Tapscott & Tapscott, 2016). Blockchain can trace a company's supply chain in real-time, enabling a clearer understanding of the environmental and social impact of production processes.

Despite its advantages, the integration of technology into sustainable finance presents challenges, particularly in terms of data privacy, accuracy, and algorithmic transparency. Ensuring that AI models remain free from bias and that data sources are ethically obtained are crucial considerations for sustainable finance. These technologies also require substantial investment, and the complex infrastructure needed may limit access for smaller firms. Nonetheless, embracing technological tools such as AI, machine learning, and blockchain could fundamentally reshape the landscape of sustainable finance, paving the way for a more data-driven and resilient approach to sustainable investment.

Emerging Tools in Finance

Emerging tools such as AI-driven ESG scoring systems, blockchain technology, and advanced data analytics are transforming sustainable finance by enhancing transparency, accuracy, and decision-making in financial markets. AI enables real-time assessment of ESG factors by processing vast datasets, including alternative sources like satellite imagery and social media, providing comprehensive and predictive insights that help financial institutions align lending terms with a company's sustainability profile. Blockchain ensures transparency and accountability in ESG reporting through immutable ledgers, mitigating concerns about greenwashing and enabling traceable verification of sustainability efforts, while smart contracts tied to ESG performance can automate adjustments to loan terms. Advanced data analytics further empowers institutions to quantify the correlation between ESG practices and financial outcomes, supporting the development of innovative financial products like green bonds and sustainability-linked loans. Despite challenges such as data standardization, implementation costs, and the environmental footprint of some technologies, these tools offer unprecedented opportunities to scale sustainable finance, driving progress toward a more resilient and inclusive financial system.

Technology, Sustainable Finance and Cost of Debt

Technology plays a pivotal role in advancing sustainable finance and its impact on the cost of debt. Innovations such as artificial intelligence (AI), machine learning, and blockchain are transforming how financial institutions integrate ESG factors into credit assessments. AI-driven algorithms enhance the ability to analyze complex ESG data, identifying trends and risks that were previously difficult to quantify. Machine learning models can predict the financial implications of sustainability initiatives, enabling lenders to adjust interest rates based on a company's ESG performance more accurately. Blockchain technology, on the other hand, promotes transparency and trust by providing immutable records of ESG-related activities, helping to counter issues like greenwashing and data inconsistencies. These advancements are particularly critical as banks increasingly rely on standardized ESG metrics to evaluate credit risk. By leveraging technology, financial institutions can streamline ESG integration, reduce information asymmetry, and offer more competitive borrowing terms to companies demonstrating strong sustainability practices. However, the adoption of such technologies also presents challenges, including the need for significant investment, regulatory oversight, and addressing concerns over data privacy and accuracy. As sustainable finance scales globally, the interplay between technology and ESG assessment will be instrumental in shaping how cost of debt evolves as a financial performance indicator.

CONCLUSIONS

This book has explored the critical transformation of finance as sustainability takes center stage, both in theory and practice. Over recent decades, the financial sector has witnessed a profound shift, with Environmental, Social, and Governance (ESG) criteria gradually embedded into regulatory frameworks, corporate strategies, and investment decisions. This shift reflects an evolving view that sustainable finance is essential not only for addressing pressing global challenges, such as climate change and social inequality, but also for enhancing long-term economic resilience and creating value for diverse stakeholders.

Key Findings on ESG Integration in Finance

The empirical analysis provided in this work show that, while ESG factors are increasingly considered in investment and corporate decision-making, their influence on financial metrics, such as the cost of debt, remains nuanced. Traditional financial criteria—profitability, risk, and creditworthiness—continue to dominate lending practices. Nevertheless, there are signs of change: sustainable companies with strong ESG performance are beginning to see lower capital costs as financial institutions and investors recognize the long-term risk mitigation these practices provide (Eccles & Klimenko, 2019). As global regulatory frameworks evolve, this effect is expected to strengthen, especially in regions like the European

Union, where initiatives like the Sustainable Finance Disclosure Regulation (SFDR) and the EU Taxonomy for Sustainable Activities aim to standardize ESG disclosures and provide a clearer structure for sustainable investments (European Commission, 2020).

Moreover, the role of ESG considerations in financial risk assessment has become evident. Climate change, for instance, introduces risks that can directly impact asset values, especially in sectors heavily reliant on natural resources. This awareness is fueling regulatory efforts, such as climate stress tests for financial institutions, aimed at protecting the financial system from unforeseen climate-related risks (Battiston et al., 2017).

PRACTICAL IMPLICATIONS FOR BANKS AND BORROWER

The findings of this book offer valuable practical implications for both banks and borrowers, providing actionable insights into the integration of ESG factors in corporate lending. For banks, the results highlight the potential to enhance credit risk assessments by incorporating ESG performance into lending decisions. ESG-aligned companies, with their focus on sustainability and resilience, are likely to present lower risk profiles, enabling banks to offer more favorable loan terms while aligning their portfolios with broader sustainability goals. This approach not only reduces financial risks associated with environmental, social, and governance challenges but also positions banks as proactive agents of sustainable development, strengthening their reputation and competitiveness in the financial market.

For borrowers, the research underscores the tangible financial benefits of adopting ESG practices. By demonstrating strong ESG performance, companies can reduce their cost of debt, thereby freeing up resources for reinvestment and growth. This reinforces the idea that sustainability is not merely a moral imperative but also a strategic advantage that can directly enhance financial performance. Borrowers can use these insights to prioritize ESG initiatives, improve transparency in reporting, and actively engage with lenders on sustainability metrics, thus fostering mutually beneficial relationships.

Moreover, the findings help bridge the gap between ESG integration and practical financial outcomes, enhancing our understanding of how sustainability influences the cost of credit. By focusing on the cost of debt, this research highlights a key financial metric that directly impacts corporate strategies and capital allocation decisions, offering a roadmap for both banks and borrowers to navigate the evolving landscape of sustainable finance. These insights reinforce the importance of aligning financial systems with sustainability principles to support long-term resilience and growth.

THE ROLE OF IMPACT INVESTING AND ETHICAL BANKING

The rise of impact investing and ethical banking models has also shown that it is possible to pursue both financial returns and social good. Impact investing, which seeks measurable social and environmental benefits, represents a growing sector that challenges the traditional profit-centric model. Ethical banks, by focusing on projects that generate societal value rather than speculative gains, provide a valuable example of how financial institutions can align their activities with sustainable development goals. These approaches illustrate that profit maximization does not have to be sacrificed when adopting a socially responsible business model, as these financial institutions are demonstrating resilience and growth potential (Bugg-Levine & Emerson, 2011).

REGULATORY DEVELOPMENTS AND THE PATH FORWARD

The regulatory landscape for sustainable finance is transforming rapidly, with Europe leading the way through policy instruments like the SFDR, the EU Taxonomy, and the Corporate Sustainability Reporting Directive (CSRD). These frameworks aim to create transparency and accountability in sustainable finance, allowing investors to make informed choices and encouraging companies to disclose non-financial metrics alongside traditional financial ones. Globally, this momentum reflects a shared understanding that sustainability is crucial to economic stability. Such regulatory measures provide a foundation for a more resilient financial system by promoting sustainability as an integral part of risk management and strategic planning (Busch, 2020).

The book's findings underscore that, while regulatory changes are fundamental to achieving a sustainable financial system, voluntary corporate initiatives and investor preferences are equally important. With these drivers, the financial sector is encouraged to adopt a multi-dimensional approach, where ESG performance is viewed not as a secondary consideration but as a cornerstone of long-term strategy.

CHALLENGES AND OPPORTUNITIES
FOR FUTURE RESEARCH

The shift toward sustainable finance presents unique challenges and opportunities. On one hand, the development of standardized metrics for ESG performance remains complex, as various sectors require different measures to assess sustainability accurately. Additionally, the risk of greenwashing—companies misrepresenting their ESG achievements to appeal to investors—highlights the need for rigorous standards and verification processes (Delmas & Burbano, 2011). Addressing these challenges will require collaborative efforts among regulators, industry leaders, and researchers to ensure that ESG metrics are reliable and meaningful.

Looking forward, there is a need for further research into the long-term financial impacts of sustainability on corporate performance and market stability. Studies exploring how ESG factors interact with other financial risks could provide a more comprehensive understanding of sustainable finance's potential for mitigating systemic risk. Moreover, the evolving role of technology, particularly artificial intelligence and big data, in ESG analysis offers a promising area for research. These tools can enhance ESG data accuracy and provide real-time insights, though ethical considerations, such as algorithmic bias, must be carefully managed (Gunning & Aha, 2019).

FINAL REFLECTIONS

In conclusion, the integration of sustainability into the financial sector represents a paradigm shift with significant implications for how companies, investors, and financial institutions operate. This transition towards a finance model that balances profit with purpose has the potential to redefine the sector's role in society, making it a driver of positive change. By incorporating ESG factors into core financial practices, the sector can contribute meaningfully to global sustainability goals, fostering economic stability and societal well-being.

As the regulatory and market landscapes continue to evolve, financial institutions must adopt a proactive approach, aligning their strategies with both ethical standards and long-term economic objectives. Sustainable finance thus offers a path forward, where financial markets can support a resilient and equitable global economy. This approach aligns with the broader societal demand for responsible corporate governance and reflects a growing consensus that, in the pursuit of profit, companies and investors must also account for the impact of their actions on people and the planet.

In this new era, sustainable finance emerges not as a trend but as a fundamental restructuring of the financial system, laying the groundwork for a future in which financial success and social responsibility are inextricably linked.

BIBLIOGRAPHY

Adams, C. A. (2015). The international integrated reporting council: A call to action. *Critical Perspectives on Accounting, 27*, 23–28.

Aguinis, H., & Glavas, A. (2012). What we know and don't know about corporate social responsibility: A review and research agenda. *Journal of Management, 38*(4), 932–968.

Albuquerque, R., Koskinen, Y., & Zhang, C. (2019). *ESG leaders often enjoy enhanced reputational benefits, which translate into reduced capital costs, including lower debt premiums.*

Attig, N., El Ghoul, S., Guedhami, O., & Suh, J. (2013). Corporate social responsibility and credit ratings. *Journal of Business Ethics, 117*(4), 679–694.

Baden, D., & Harwood, I. A. (2013). Terminology matters: A critical exploration of corporate social responsibility terms. *Journal of Business Ethics, 116*(3), 615–627.

Baker, M. (2003). *Corporate social responsibility—What does it mean?* Retrieved from www.mallenbaker.net

Banerjee, S. B., & Jackson, L. (2017). Microfinance and the sustainable development goals: A critical assessment. *Development Policy Review, 35*(4), 469–486.

Banga, J. (2019). The green bond market: A potential source of climate finance for developing countries. *Journal of Sustainable Finance & Investment, 9*(1), 17–32.

Batten, S., Sowerbutts, R., & Tanaka, M. (2016). *Let's talk about the weather: The impact of climate change on central banks* (Bank of England Staff Working Paper No. 603).

Battiston, S., Mandel, A., Monasterolo, I., Schütze, F., & Visentin, G. (2017). A climate stress-test of the financial system. *Nature Climate Change, 7*(4), 283–288.

Bauer, R., & Hann, D. (2010). *Corporate environmental management and credit risk* (Working Paper). Maastricht University.

Benn, S., & Bolton, D. (2011). *Key concepts in corporate social responsibility.* Sage.

Berg, F., Kölbel, J. F., & Rigobon, R. (2020). *The alignment of lending practices with global initiatives, such as the Paris Agreement, reinforces the role of banks as critical drivers of sustainable development.*

Bollen, J., Mao, H., & Zeng, X. (2011). Twitter mood predicts the stock market. *Journal of Computational Science, 2*(1), 1–8.

Bolton, P., Després, M., Pereira Da Silva, L. A., Samama, F., & Svartzman, R. (2020). *The green swan: Central banking and financial stability in the age of climate change.* Bank for International Settlements.

Bowman, W. (2017). Measuring nonprofit impact. *Nonprofit Management and Leadership, 27*(3), 373–384.

Brest, P., & Born, K. (2013). When can impact investing create real impact? *Stanford Social Innovation Review, 11*(4), 22–31.

Bugg-Levine, A., & Emerson, J. (2011). Impact investing: Transforming how we make money while making a difference. *Innovations, 6*(3), 9–18.

Busch, D. (2020). Sustainable finance regulation and its challenges. *European Business Law Review, 31*(4), 607–631.

Cade, N. L. (2018). Corporate social media: How two-way disclosure channels influence investors. *Accounting, Organizations and Society, 68*, 63–79.

Caldecott, B., Howarth, N., & McSharry, P. (2015). *Stranded assets in agriculture: Protecting value from environment-related risks.* Oxford Smith School of Enterprise and the Environment.

Campiglio, E. (2016). Beyond carbon pricing: The role of banking and monetary policy in financing the transition to a low-carbon economy. *Ecological Economics, 121*, 220–230.

Carney, M. (2021). *Value(s): Building a better world for all.* PublicAffairs.

Carroll, A. B. (1979). A three-dimensional conceptual model of corporate performance. *Academy of Management Review, 4*(4), 497–505.

Carroll, A. B. (1991). The pyramid of corporate social responsibility: Toward the moral management of organizational stakeholders. *Business Horizons, 34*(4), 39–48.

Carroll, A. B., & Buchholtz, A. K. (2008). *Business and society: Ethics, sustainability, and stakeholder management.* Cengage Learning.

Carroll, A. B., & Shabana, K. M. (2010). The business case for corporate social responsibility: A review of concepts, research and practice. *International Journal of Management Reviews, 12*(1), 85–105.

Chava, S. (2014). Environmental externalities and cost of capital. *Management Science, 60*(9), 2223–2247.

Chen, Z., Luo, Z., & Wu, X. (2020). Financial technology and sustainable development: Opportunities and challenges. *Journal of Cleaner Production, 276,* 124–132.

Cheng, B., Ioannou, I., & Serafeim, G. (2014). *Larger firms or those in heavily regulated industries may experience greater scrutiny of their ESG practices, resulting in more pronounced effects on loan terms.*

Clark, G. L., Feiner, A., & Viehs, M. (2015). *From the stockholder to the stakeholder: How sustainability can drive financial outperformance.* University of Oxford and Arabesque Partners.

Clarkson, M. E. (1995). A stakeholder framework for analyzing and evaluating corporate social performance. *Academy of Management Review, 20*(1), 92–117.

Climate Bonds Initiative. (2021). *EU Green Bond Standard.* www.climatebonds.net

Cooper, T., & Uzur, D. (2015). Sustainability and bank loan interest rates. *Journal of Corporate Finance, 29,* 342–356.

Crotty, J. (2009). Structural causes of the global financial crisis: A critical assessment of the 'new financial architecture.' *Cambridge Journal of Economics, 33*(4), 563–580.

D'Amato, D., Droste, N., Allen, B., Kettunen, M., Lahtinen, K., Korhonen, J., Leskinen, P., Matthies, B. D., & Toppinen, A. (2019). Green, circular, bio economy: A comparative analysis of sustainability avenues. *Journal of Cleaner Production, 168,* 716–734.

Dahlsrud, A. (2008). How corporate social responsibility is defined: An analysis of 37 definitions. *Corporate Social Responsibility and Environmental Management, 15*(1), 1–13.

de Jesus, A., & Mendonça, S. (2018). Lost in transition? Drivers and barriers in the eco-innovation road to the circular economy. *Ecological Economics, 145,* 75–89.

de Sousa, D. M. (2021). The EU's sustainable finance strategy: Continuity or change? *Journal of European Integration, 43*(2), 211–226.

Delmas, M. A., & Burbano, V. C. (2011). The drivers of greenwashing. *California Management Review, 54*(1), 64–87.

Deng, X., Kang, J. K., & Low, B. S. (2013). Corporate social responsibility and stakeholder value maximization: Evidence from mergers. *Journal of Financial Economics, 110*(1), 87–109.

Devalle, A., Fiandrino, S., & Cantino, V. (2017). The link between ESG and financial performance in Europe. *Journal of Applied Accounting Research, 18*(1), 120–145.

Di Tullio, P., & Pascale, P. (2020). The impact of EU directive 2014/95 on corporate non-financial disclosure: The case of Italy. *Journal of Business Ethics, 161*(4), 803–821.

Ding, R., & Li, H. (2020). Artificial intelligence in green finance: Mechanisms and applications. *Environmental Research Letters, 15*(5), 540–555.

Donaldson, T., & Preston, L. E. (1995). The stakeholder theory of the corporation: Concepts, evidence, and implications. *Academy of Management Review, 20*(1), 65–91.

Donna, G. (2021). *Ethics and economic theory: Towards a modern approach.* Routledge.

Eccles, R. G., & Klimenko, S. (2019). The investor revolution. *Harvard Business Review, 97*(3), 106–116.

Eccles, R. G., & Krzus, M. P. (2010). *One report: Integrated reporting for a sustainable strategy.* Wiley.

Eccles, R. G., & Serafeim, G. (2013). The performance frontier: Innovating for a sustainable strategy. *Harvard Business Review, 91*(5), 50–60.

Eccles, R. G., Ioannou, I., & Serafeim, G. (2014). The impact of corporate sustainability on organizational processes and performance. *Management Science, 60*(11), 2835–2857.

Elkington, J. (1997). *Cannibals with forks: The triple bottom line of 21st century business.* Capstone Publishing.

Elkington, J. (2018). *25 years ago I coined the phrase "triple bottom line." Here's why it's time to rethink it.* Harvard Business Review.

European Commission Technical Expert Group on Sustainable Finance. (2019). *Taxonomy Technical Report.*

European Commission. (2014). Directive 2014/95/EU of the European Parliament and of the Council. *Official Journal of the European Union.*

European Commission. (2019). *EU taxonomy for sustainable finance.* European Union.

European Commission. (2019). The European Green Deal.

European Commission. (2020). *A new Circular Economy Action Plan for a cleaner and more competitive Europe.*

European Commission. (2020). *Sustainable Finance Disclosure Regulation (SFDR).*

European Commission. (2021). *Proposal for a Corporate Sustainability Reporting Directive (CSRD).*

European Commission. (2021). *The European Green Deal.* European Union.

European Financial Reporting Advisory Group (EFRAG). (2021). *Project task force on preparatory work for the elaboration of possible EU non-financial reporting standards.*

European Union. (2019). *Regulation (EU) 2019/2088 of the European Parliament and of the Council on sustainability-related disclosures in the financial services sector.*

Flammer, C. (2021). Corporate green bonds. *Journal of Financial Economics, 142*(2), 499–516.

Freeman, R. E. (1984). *Strategic management: A stakeholder approach.* Cambridge University Press.

Freeman, R. E., & Reed, D. L. (1983). Stockholders and stakeholders: A new perspective on corporate governance. *California Management Review, 25*(3), 88–106.

Freeman, R. E., Harrison, J. S., & Wicks, A. C. (2004). *Modern theories of stakeholder capitalism emphasize that firms managing ESG issues effectively create long-term value for both shareholders and creditors.*

Freeman, R. E., Harrison, J. S., Wicks, A. C., Parmar, B. L., & de Colle, S. (2007). *Stakeholder theory: The state of the art.* Cambridge University Press.

Friede, G., Busch, T., & Bassen, A. (2015). *ESG factors are generally associated with superior financial performance, reinforcing the argument that sustainability reduces risk exposure.*

Friede, G., Busch, T., & Bassen, A. (2015). ESG and financial performance: Aggregated evidence from more than 2000 empirical studies. *Journal of Sustainable Finance & Investment, 5*(4), 210–233.

Friedman, M. (1962). *Capitalism and freedom.* University of Chicago Press.

Friedman, M. (1970, September 13). The social responsibility of business is to increase its profits. *New York Times Magazine.*

Frost, T. S., Hodgson, A., & Seefried, K. (2019). Banking on a greener future: A framework for sustainable finance. *Journal of Sustainable Finance & Investment, 9*(4), 287–302.

Gangi, F., Meles, A., Daniele, L. M., Varrone, N., & Salerno, D. (2021). *The evolution of sustainable investments and finance: Theoretical perspectives and new challenges.* Springer Nature.

Gao, M., & Geng, X. (2024). The role of ESG performance during times of COVID-19 pandemic. *Scientific Reports, 14*(1), 2553.

Ge, W., & Liu, M. (2015). Corporate social responsibility and the cost of corporate bonds. *Journal of Accounting and Public Policy, 34*(6), 597–624.

Geissdoerfer, M., Savaget, P., Bocken, N. M., & Hultink, E. J. (2017). The circular economy–A new sustainability paradigm? *Journal of Cleaner Production, 143*, 757–768.

GIIN. (2021). *Annual impact investor survey.* Global Impact Investing Network.

Goss, A., & Roberts, G. S. (2011). The impact of corporate social responsibility on the cost of bank loans. *Journal of Banking & Finance, 35*(7), 1794–1810.

GRI. (2011). *Global reporting initiative. Sustainability reporting guidelines.* GRI.

Gunning, D., & Aha, D. (2019). DARPA's explainable artificial intelligence program. *AI Magazine, 40*(2), 44–58.

Harrison, J. S., & Freeman, R. E. (1999). Stakeholders, social responsibility, and performance: Empirical evidence and theoretical perspectives. *Academy of Management Journal, 42*(5), 479–485.

Hasan, I., Hoi, C. K., Wu, Q., & Zhang, H. (2022). Does social capital matter for bank loan pricing? *Journal of Corporate Finance, 72*, 102151.

HLEG (High-Level Expert Group on Sustainable Finance). (2018). *Financing a sustainable European economy*. European Commission.

Hoepner, A. G., Oikonomou, I., Scholtens, B., & Schröder, M. (2016). The effects of corporate and country sustainability characteristics on the cost of debt: An international investigation. *Journal of Business Finance & Accounting, 43*(1–2), 158–190.

Huang, T., Krasa, S., & Villamil, A. (2020). The impact of information acquisition on corporate decisions. *American Economic Journal: Macroeconomics, 12*(3), 1–26.

IFRS Foundation. (2021). *IFRS Foundation Trustees Announce Sustainability Standards Board.*

Ioannou, I., & Serafeim, G. (2019). *Corporate sustainability: A strategy?* Harvard Business Review.

Jackson, E. T. (2013). Interrogating the theory of change: Evaluating impact investing where it matters most. *Journal of Sustainable Finance & Investment, 3*(2), 95–110.

Jensen, M. C., & Meckling, W. H. (1976). Theory of the firm: Managerial behavior, agency costs and ownership structure. *Journal of Financial Economics, 3*(4), 305–360.

Jiang, Y., & Muradoğlu, Y. (2021). ESG and the cost of bank loans. *Journal of Banking & Finance, 130*, 106–115.

Jones, T. M. (1980). Corporate social responsibility revisited. *Redefined. California Management Review, 22*(2), 59–67.

Jones, T. M. (1995). Instrumental stakeholder theory: A synthesis of ethics and economics. *Academy of Management Review, 20*(2), 404–437.

Jones, T. M., Wicks, A. C., & Freeman, R. E. (2002). Stakeholder theory: The state of the art. In *The Blackwell guide to business ethics* (pp. 19–37). Blackwell.

Kanter, R. M. (2011). How great companies think differently. *Harvard Business Review, 89*(11), 66–78.

Khan, M., Serafeim, G., & Yoon, A. (2016). Corporate sustainability: First evidence on materiality. *The Accounting Review, 91*(6), 1697–1724.

Kohlscheen, E. (2021). The impact of COVID-19 on interest rates and policy decisions in Europe. *European Economic Review, 133*, 103669.

Korhonen, J., Nuur, C., Feldmann, A., & Birkie, S. E. (2018). Circular economy as an essentially contested concept. *Journal of Cleaner Production, 175*, 544–552.

Kotsantonis, S., Pinney, C., & Serafeim, G. (2016). ESG integration in investment management: Myths and realities. *Journal of Applied Corporate Finance, 28*(2), 10–16.

KPMG. (2021). *SEC's Potential Climate Disclosure Requirements*.

Kraus, S., Rehman, S. U., & García, F. J. S. (2020). Corporate social responsibility and environmental performance: The mediating role of environmental strategy and green innovation. *Technological Forecasting and Social Change, 160*, 120262.

Lash, J., & Wellington, F. (2007). Competitive advantage on a warming planet. *Harvard Business Review, 85*(3), 94–102.

Lee, M. D. P. (2008). A review of the theories of corporate social responsibility: Its evolutionary path and the road ahead. *International Journal of Management Reviews, 10*(1), 53–73.

Liang, H., & Renneboog, L. (2017). On the foundations of corporate social responsibility. *Journal of Finance, 72*(2), 853–910.

Martin, M., & Gregory, R. (2015). *Scaling impact: Blueprint for collective action to scale impact investment in and from the UK*. Social Finance UK.

Mauck, N., & Salzsieder, L. (2018). Financial crisis, regulatory response, and the impact on systemic risk: Evidence from the U.S. *Journal of Banking & Finance, 87*, 153–165.

Meadows, D. H., Meadows, D. L., Randers, J., & Behrens, W. W. III. (1972). *The limits to growth*. Potomac Associates.

Mehrabi, N., Morstatter, F., Saxena, N., Lerman, K., & Galstyan, A. (2021). A survey on bias and fairness in machine learning. *ACM Computing Surveys (CSUR), 54*(6), 1–35.

Mitchell, R. K., Agle, B. R., & Wood, D. J. (1997). Toward a theory of stakeholder identification and salience: Defining the principle of who and what really counts. *Academy of Management Review, 22*(4), 853–886.

Monasterolo, I., & de Angelis, L. (2020). Are financial markets pricing carbon risks after the Paris Agreement? *Review of Financial Economics, 38*(3), 452–475.

Moser, D. V., & Martin, P. R. (2012). A broader perspective on corporate social responsibility research in accounting. *The Accounting Review, 87*(3), 797–806.

Nandy, M., & Lodh, S. (2012). Do banks value the eco-friendliness of firms in their corporate lending decision? Some empirical evidence. *International Review of Financial Analysis, 25*, 83–93.

Ness, D. (2008). Sustainable urban infrastructure in China: Towards a factor 10 improvement in resource productivity through integrated infrastructure systems. *International Journal of Sustainable Development & World Ecology,* *15*(4), 288–301.

Network for Greening the Financial System (NGFS). (2019). *A call for action: Climate change as a source of financial risk.*

Oikonomou, I., Brooks, C., & Pavelin, S. (2014) "ESG-related controversies can increase firms' credit spreads, particularly when governance issues are perceived as a significant risk factor."

Palmieri, E., & Geretto, E. F. (2024). *The determinants of bank-firm relationships: Overview, evolution, and challenges. Adapting to change: ESG and alternative finance in shaping the bank-firm relationship.* Palgrave Macmillan Studies in Banking and Financial Institutions.

People's Bank of China. (2020). *China's Green Bond Guidelines.*

Pérez, A., Martínez, P., & del Bosque, I. R. (2021). The role of customer CSR expectations in motivating customer citizenship behavior. *Journal of Business Research, 129,* 104–112.

Pfitzer, M., Bockstette, V., & Stamp, M. (2013). Innovating for shared value. *Harvard Business Review, 91*(9), 100–107.

Pizzi, S., Caputo, A., & Venturelli, A. (2021). ESG reporting practices and the EU regulation on non-financial information: The experience of Italian listed companies. *Sustainability, 13*(8), 3660.

Porter, M. E., & Kramer, M. R. (2011). Creating shared value: How to reinvent capitalism—And unleash a wave of innovation and growth. *Harvard Business Review, 89*(1/2), 62–77.

Porter, M. E., Hills, G., Pfitzer, M., Patscheke, S., & Hawkins, E. (2012). *Measuring shared value: How to unlock value by linking social and business results.* FSG.

Prieto-Sandoval, V., Jaca, C., & Ormazabal, M. (2018). Towards a consensus on the circular economy. *Journal of Cleaner Production, 179,* 605–615.

Raji, I. D., & Buolamwini, J. (2020). Actionable auditing: Investigating the impact of publicly naming biased performance results of commercial AI products. In *Proceedings of the 2020 Conference on Fairness, Accountability, and Transparency* (pp. 44–56).

Rockström, J., Steffen, W., Noone, K., Persson, Å., Chapin, F. S., III, Lambin, E., Lenton, T. M., Scheffer, M., Folke, C., Joachim Schellnhuber, H., Nykvist, B., de Wit, C. A., Hughes, T., van der Leeuw, S., Rodhe, H., Sörlin, S., Snyder, P. K., Costanza, R., Svedin, U., Falkenmark, M., ... & Foley, J. A. (2009). A safe operating space for humanity. *Nature, 461*(7263), 472–475.

Sarkar, S., & Searcy, C. (2016). Zeitgeist or chameleon? A quantitative analysis of CSR definitions. *Journal of Cleaner Production, 135,* 1423–1435.

Schoenmaker, D., & Schramade, W. (2019). *Principles of sustainable finance.* Oxford University Press.

Serafeim, G. (2020). Social-impact efforts that create real value. *Harvard Business Review, 98*(5), 38–48.

Shiller, R. J. (2017). Narrative economics. *American Economic Review, 107*(4), 967–1004.

Soppe, A. (2009). Sustainable finance as a connection between corporate social responsibility and social responsible investing. *European Business Review, 21*(2), 207–223.

Stern, D. I. (2012). The role of energy in economic growth. *Annals of the New York Academy of Sciences, 1219*(1), 26–51.

Sullivan, R., & Gouldson, A. (2020). *The governance of climate change: Political and policy challenges.*

Tang, X., & Demeritt, D. (2020). Blockchain technology for environmental sustainability: A systematic review. *Journal of Environmental Management, 256*, 109–126.

Tapscott, D., & Tapscott, A. (2016). *Blockchain revolution: How the technology behind bitcoin is changing money, business, and the world.* Penguin.

Turner, G. (2008). A comparison of the limits to growth with 30 years of reality. *Global Environmental Change, 18*(3), 397–411.

UN. (2015). *Transforming our world: The 2030 Agenda for Sustainable Development.* United Nations.

UNEP FI. (2019). *The Principles for Responsible Banking established by the United Nations Environment Programme Finance Initiative reflect the growing trend of integrating non-financial metrics into credit risk evaluation.*

UNEP. (2019). *United Nations environment programme: Promoting environmental sustainability.* United Nations Environmental Programme.

Wang, H., Cao, Y., & Li, X. (2021). Blockchain-based solutions in the green supply chain: Benefits and limitations. *International Journal of Production Research, 59*(3), 779–791.

Wood, D. J. (1991). Corporate social performance revisited. *Academy of Management Review, 16*(4), 691–718.

Yale Center for Environmental Law & Policy. (2022). *2022 environmental performance index.*

Yang, J., & Li, F. (2021). Green fintech: AI and big data analytics in sustainable finance. *Sustainability, 13*(7), 4012.

Zhou, L., & Tian, M. (2019). Mobile finance and sustainability: Assessing the rise of green digital banking. *Global Environmental Change, 58*, 126–137.

Index

The manufacturer's authorised representative in the EU is Springer
Nature Customer Service Centre GmbH, Europaplatz 3, 69115 Heidelberg,
Germany. If you have any concerns regarding our products, please
contact ProductSafety@springernature.com

Printed and bound by CPI Group (UK) Ltd, Croydon, CR0 4YY

24/04/2026

02096371-0002